Practical Antenna

Antennas are an important component of every communication systems, including radio, television, satellite, radar, and cellular networks which transmit and receive radio waves over the wireless channel. The field of antenna is quite vast, and with the current development in the wireless technologies enormous amount of work and effort has been put in the design, analysis, and applications of antennas. As an individual, a reader has to put a lot of effort to understand the modern antenna analysis and design for various applications. Therefore, the editors and authors of this book have come up with an idea of putting all the necessary information required at one centralized place in the form of a book. The readers of this book will be able to access all the necessary information of antennas, from basic to advanced, theory to practical and its modern applications. The text provides information on antenna design for next-generation communication systems and Internet of Things (IoT) applications, followed by the integration of the antennas in the wireless system, and also covers multi-input and multi-output (MIMO) antennas, metasurface antennas, reconfigurable antennas, fractal antennas, and design of beamforming networks.

This book:

- Focuses on modern antennas for 5G communication systems and next-generation IoT networks
- Highlights fractal radiator-based printed linear antenna arrays designed for direct broadcast satellite applications with a particular focus on direct-to home (DTH) services
- Covers the implementation of modern swarm intelligence optimization techniques for antenna array pattern synthesis
- Includes general framework for the design and optimization of metasurface antennas
- Presents design and development of high-gain millimeter-wave beam switching antennas using passive frequency selective surfaces (FSS)

This book can also serve as a valuable reference for undergraduate, postgraduate students, as well as researchers working in radio frequency (RF) and microwave.

Smart Technologies for Engineers and Scientists
Series Editor: Mangey Ram

Applications of Mathematical Modeling, Machine Learning, and Intelligent Computing for Industrial Development
Madhu Jain, Dinesh K. Sharma, Rakhee Kulshrestha, and H.S. Hota

Practical Antenna

Design, Analysis, and Applications

Edited by

Abhishek Kumar Awasthi, Abhinav Sharma, and Kushmanda Saurav

CRC Press
Taylor & Francis Group
Boca Raton London New York

CRC Press is an imprint of the
Taylor & Francis Group, an **informa** business

Designed cover image: shutterstock

First edition published 2025
by CRC Press
2385 NW Executive Center Drive, Suite 320, Boca Raton FL 33431

and by CRC Press
4 Park Square, Milton Park, Abingdon, Oxon, OX14 4RN

CRC Press is an imprint of Taylor & Francis Group, LLC

ISBN: 978-1-032-74366-0 (hbk)
ISBN: 978-1-032-74780-4 (pbk)
ISBN: 978-1-003-47085-4 (ebk)

DOI: 10.1201/9781003470854

Typeset in Sabon
by Newgen Publishing UK

Contents

Preface

Antennas are an integral part of any wireless link and with the current development in the wireless technologies, there is a rapid growth in the field of antenna design. Some major wireless applications are next-generation communication systems, remote sensing, body-centric network, Internet of Things (IoT), vehicular communication, and automotive radars. The focus of this book is to give readers all the fundamental concepts and diverse applications of antennas at one centralized place. The organization of the book is as follows:

Chapter 1 presents an innovative methodology for the design of a dual-band circularly polarized (CP) multiple-input–multiple-output (MIMO) antenna, incorporating a dual-band linear-to-circular polarization converter. The converter transforms an incoming linearly polarized (LP) wave into a CP wave with orthogonal polarization at the lower- and upper-frequency bands of operation.

Chapter 2 provides a general framework for the design and optimization of metasurface antennas, offering a comprehensive exploration of key principles, methodologies, and recent advancements. Practical considerations, such as measurement and testing methodologies, are presented to guide researchers and engineers in evaluating the performance of metasurface antennas.

Chapter 3 reviews and provides an in-depth analysis of CP antennas and techniques to achieve wideband and multiband characteristics. The chapter also presents different techniques to generate circular polarization and methods to achieve bandwidth enhancement, and multiband response.

Chapter 4 provides a birds-eye view of transformative impact of MIMO reconfigurable antennas on wireless communication systems. The chapter also presents a case study which highlights the design and optimization of reconfigurable MIMO antenna for IoT applications.

Chapter 5 examines the radiation mechanism of a single element dipole antenna, elucidating its functioning, polarization properties, and radiation pattern influenced by antenna design. It explores different factors affecting the radiation mechanism such as antenna length, operating frequency, and environmental conditions, highlighting their impact on radiation efficiency and pattern.

Chapter 6 presents methodology for the design of a four-element millimeter-wave MIMO antenna for 5G wireless communication systems. The performance of the MIMO antenna is enhanced by modifying the ground plane with a diagonal square slot defected ground structure.

Chapter 7 explores sooty tern optimization (STO) algorithm for pattern synthesis of linear antenna array. The amplitude of the excitation current is optimized for peak side lobe level minimization, close-in side lobe level reduction and null placement in the specified directions.

Chapter 8 presents design and development of 1-D and 2-D high-gain millimeter wave beam switching antennas using passive frequency selective surfaces (FSS)-based transmitarrays. The 1-D and 2-D switchable beam transmitarray antennas is used for beamforming in base stations and vehicular networks for future 5G millimeter wave communications.

Chapter 9 proposes a pair of fractal radiator-based printed linear antenna arrays designed specifically for direct broadcast satellite applications with a particular focus on direct-to-home (DTH) services. The Minkowski fractal technique is used to create the basic framework of the proposed design and to enhance the effective fractal length, the proposed fractal structure has been positioned in the four quadrants which generate a symmetrical radiation pattern in the required direction.

Chapter 10 proposes a systematic approach to design a microstrip patch array antenna for IoT applications. Two-element and four-element microstrip patch array antennas are designed by duplicating the single microstrip antenna to improve the gain and bandwidth with a quarter wavelength distance using a software program with specific features.

The readers of this book will be able to access all the necessary information, from basic to advanced, theory to practical approach of antennas. This book covers the basic understanding of antennas along with its advanced application such as 5G communication, IoT, and DTH services. This book serves as a valuable reference and has information for almost all the class of readers such as undergraduate or post-graduate students, researchers from different fields, domain experts, industrialists, academicians, etc.

Acknowledgments

We acknowledge CRC Press – Taylor & Francis Group for this opportunity and professional support. Our special thanks to Gagandeep Singh, Publisher Engineering, CRC Press, Gauravjeet Singh Reen, Senior Editor – Engineering, and Isha Ahuja, Editorial Assistant – Engineering for the excellent support provided us to complete this book.

Thanks to the chapter authors and reviewers for their availability for this work.

Abhishek Kumar Awasthi, Paras Antidrone Technologies Pvt. Ltd., India
Abhinav Sharma, University of Petroleum and Energy Studies, India
Kushmanda Saurav, IIT Jammu, India

Editors

Abhishek Kumar Awasthi completed his B.Tech. in Electronics and Communication Engineering from Ajay Kumar Garg Engineering College Ghaziabad (UPTU Lucknow) in the year 2008. He did his M. Tech. in Digital Communication from Ambedkar Institute of Technology Delhi (GGSIP University Delhi) in 2011 and doctoral degree from the Department of Electrical Engineering (RF and Microwave) of Indian Institute of Technology Kanpur in 2018. Currently, he is working as a Deputy Manager RF, Paras Antidrone Technologies Private Limited, Mumbai, India. Prior to that he has worked at Remote Sensing Center at The University of Alabama, Tuscaloosa, AL, USA, as an associate research engineer. His major research interests include design, optimize, fabricate, and develop various active and passive microwave circuits for the radars. Particularly, he is working on the radars with application in ice sounding and soil moisture measurement. He has also actively worked on the assembly and field test of the radar for soil moisture and snow measurements. Apart from radar design, he has worked on the antenna array design for various applications. He has mainly worked on design of the antenna arrays which includes the study of inter-element coupling and its effects on the performance of the antenna arrays, design of compact and ultra-wide band antenna arrays, shared aperture antenna arrays, and antenna arrays for airborne applications.

Abhinav Sharma received his B.Tech. from H.N.B. Garhwal University, Srinagar, India, in 2009 and the M.Tech. and Ph.D. from Govind Ballabh Pant University of Agriculture and Technology, Pantnagar, India, in 2011 and 2016. He is working as an associate professor in the Department of Electrical and Electronics Engineering in University of Petroleum and Energy Studies, Dehradun. He has taught subjects such as data communication networks, microprocessor and embedded system, artificial intelligence and machine learning, and communication systems. He has publications in many national and international journals and conferences. His field of interest includes adaptive array signal processing, artificial intelligence, and machine learning, optimization techniques, and smart antennas.

Kushmanda Saurav received the B.E. degree in Electronics and Telecommunication from Bengal Engineering and Science University, Shibpur, India, in 2009, the M.Tech. degree in Electronics and Communication Engineering from IIT Roorkee, India, in 2011, and the Ph.D. degree in Electrical Engineering from IIT Kanpur, India, in 2017. He worked as a postdoctoral fellow with the Royal Military College Canada. He is currently an assistant professor with the Department of Electrical Engineering, IIT Jammu, India. His research interests include microwave and millimeter wave antennas and frequency selective surfaces. He has published more than 75 research papers in esteemed journals and conferences. He has served as a reviewer in IEEE, IET, and Wiley Journals. He is actively involved in the professional activities of the IEEE. He is a senior member of IEEE and currently the faculty advisor of IEEE MTTS and APS Student Branch Chapters of IIT Jammu. He has rich experience of supervising M.Tech. and Ph.D. students and undertaking sponsored projects.

Contributors

P. S. Aithal
Srinivas University Mangalore,
Karnataka, India

Pooja Anand
Amity University Greater Noida,
Uttar Pradesh, India

Abhishek Kumar Awasthi
Paras Antidrone Technologies Private Limited, Navi Mumbai,
Maharashtra, India

Somak Bhattacharyya
Indian Institute of Technology BHU, Varanasi,
Uttar Pradesh, India

Anirban Chatterjee
National Institute of Technology Goa, Farmagudi, Ponda,
Goa, India

Vartika Dahima
University of Petroleum & Energy Studies, Dehradun,
Uttarakhand, India

Lipsa Das
Amity University Greater Noida,
Uttar Pradesh, India

Javid Ahmad Ganie
Indian Institute of Technology Jammu,
Jammu & Kashmir, India

Kalyan Sundar Kola
Department of CSE, Brainware University, Barasat,
Kolkata, West Bengal, India

Shiban Kishen Koul
Indian Institute of Technology Delhi,
New Delhi, India

N. Dinesh Kumar
Vignan Institute of Technology and Science Hyderabad,
Telangana, India

Trivesh Kumar
Indian Institute of Information Technology, Design and Manufacturing,
Jabalpur, Madhya Pradesh, India

Ranjan Mishra
University of Petroleum & Energy Studies, Dehradun,
Uttarakhand, India

Anoop Raghuvanshi
Birla Institute of Applied Sciences, Bhimtal,
Nainital, Uttarakhand, India

Diptiranjan Samantaray
University of Delhi,
New Delhi, India

Kapil Saraswat
Central University of Rajasthan, Ajmer,
Rajasthan, India

Kushmanda Saurav
Indian Institute of Technology Jammu,
Jammu & Kashmir, India

Abhinav Sharma
University of Petroleum & Energy Studies, Dehradun,
Uttarakhand, India

F. B. Shiddanagouda
Srinivas University Mangalore,
Karnataka, India

Sarala S. Shirabadagi
RV Institute of Technology and Management Bangalore,
Karnataka, India

Akanksha Singh
Amity University, Greater Noida,
Uttar Pradesh, India

Sunaina Singh
University of Petroleum & Energy Studies, Dehradun,
Uttarakhand, India

Mohammad Ayoub Sofi
Nanyang Technological University Singapore,
Singapore

R. M. Vani
Gulbarga University Kalaburagi,
Karnataka, India

Chapter 1

Dual-band circularly polarized MIMO antenna deploying a dual-band transmission-type linear to circular polarization converter

Mohammad Ayoub Sofi, Kushmanda Saurav, and Shiban Kishen Koul

1.1 INTRODUCTION

An electromagnetic (EM) wave represents the synchronized oscillations of electric and magnetic fields with respect to time and position traveling through the vacuum medium at the velocity equivalent to that of light. Apart from the parameters of the EM wave such as frequency, wavelength, speed, energy, and momentum, polarization has a significant impact. The temporal advancement of the orientation and relative amplitude of electric field vector in an EM wave at a fixed location in space is described by the polarization [1]. Polarization plays a significant part in several applications like optical imaging, contrast imaging, optical sensing [2], and telecommunication [3].

The polarization of EM waves can be altered using typical methods including the use of birefringent material like quartz crystal [4] and Faraday effect [5]. The typical methods employed to carry out polarization conversion are narrow band, use bulkier materials, and also attain polarization conversion at particular wavelengths. Furthermore, polarization conversion using these techniques is heavily dependent on the incidence angle of the incoming wave. So, smaller bandwidth and angle of incidence dependence make these conventional methods less suitable in modern polarization control devices [6]. To overcome these limitations, microwave engineers and scientists have come up with structures called frequency selective surfaces (FSS) [7], [8], and metasurfaces [9], [10], which can control or alter the polarization state of EM waves.

For satellite communication (Sat-Com), achieving improved isolation between transmitted and received signals necessitates the operation of an antenna across two frequency bands that are nonadjacent, while also employing orthogonal polarization. Hence, dual-band antennas with the ability to produce orthogonal circular polarization (CP) waves present a viable avenue for investigation. In existing literature, various antennas with CP configurations are reported, including slotted antennas [11],

DOI: 10.1201/9781003470854-1

cavity-supported crossed dipoles [12], consecutively turned antennas [13], and layered patch antennas [14]. Within the realm of Sat-Com, there is considerable research dedicated to Ka-band antennas to enable high-speed data links, leveraging their connection to elevated frequencies. The circularly polarized multiple-input–multiple output (MIMO) antipodal tapered slot antenna designed for C-band frequency is described in detail in reference [15]. A unique excitation system incorporating couplers, crossovers, and some lines for delay in an MIMO to achieve broadband performance is employed which makes the antenna's polarization diverse. The double-frequency operating CP MIMO antenna described in reference [16] is intended to function in two different 5G channels. This antenna configuration highlights a nonuniform slot antenna with an elliptical shape, achieving dual-band performance and CP characteristics by employing two asymmetrical L-shaped elements within the excitation system. Reference [17] presents an MIMO antenna using dielectric resonating element rectangular in shape for 5G applications operating in dual bands and exhibiting CP characteristics. The antenna design employs L- and C-like openings within the bottom surface to induce excitation. An L-shaped opening triggers the first resonance, while C-like opening causes the other resonance. Reference [18] details a mono-layered metasurface integrated antenna with CP. The antenna prints its metasurface cells and antenna onto the same dielectric, resulting in a simple design. A CP MIMO antenna with large bandwidth designed specifically for biotelemetry is reported in reference [19]. The antenna configuration employs four microstrip antennas positioned orthogonally to each other. The phase adjustment in reference [20] among the orthogonal modes is controlled through a metallic element inside the bottom surface inductive in nature. This approach facilitates the creation of circularly polarized wave. In reference [21], a dielectric resonator CP MIMO antenna is excited employing a step impedance resonator. Additionally in reference [22], a two-port MIMO antenna using dielectric resonators attains CP via its element geometry, positioning the two units together. This approach yields the perpendicular modes with axial ratio (AR) less than 3 dB at 7.88 GHz. Reference [23] introduces a four-port MIMO antenna suited to below –6 GHz uses. The antenna provides the CP characteristics, attained via slotting the ground integrated with a feedline.

Reference [24] discusses a multiport MIMO dielectric resonating antenna that utilizes a round epsilon-shape structure. The antenna stimulates the hybrid modes and realizes the pattern diversity. In reference [25], a planar inverted-F antenna realizes both polarization and spatial diversity through the diagonal arrangement of its element structures. The majority of the circularly polarized antennas with MIMO properties discussed in the literature operate either at a single frequency band [15], [18]–[20], [22], [23], and [25]–[28], or they support dual-band operations [16], [17], [21], with consistent polarization orientation across both frequency bands. The current study proposes a circularly polarized antenna where the polarization at the orthogonal ports can be changed. This chapter presents a method for creating a dual-band circularly polarized MIMO antenna by incorporating a converter from linear polarization (LP) to CP in dual bands [29], [30] operating in transmission mode. This converter changes an LP wave entering it into a CP wave, ensuring orthogonal polarization in operating frequency. By positioning the four MIMO components perpendicular to one another, polarization and spatial diversity are accomplished. As far as the author is aware, this study reports the foremost MIMO antenna that can operate in orthogonal CP over two different frequency bands, while also combining spatial diversity and polarization.

In contrast to conventional methods that entail complex feed networks like substrate embedded feeding [31], sequentially turned around integrated feeding [32], feeding gap waveguide [33], inverse microstrip waveguide [34], and feed layouts with ground connections [35], presented approach to design MIMO radiating structure with CP seems to be simpler as well as more direct. The presented MIMO designing idea is particularly appropriate for fixed satellite services frequency bands [29].

1.2 DUAL-BAND LINEAR TO CIRCULAR POLARIZATION CONVERTER: DESIGN AND CHARACTERIZATION

1.2.1 Conceptualization and operational principles

Figure 1.1 illustrates the suggested converter's topology. The transmitted wave is composed of both x- and y-polarized components when it is impinged upon by a plane wave. The orthogonal transmission components in the suggested converter show almost the same magnitudes in the lower and upper frequency regions of operation, with phase variances of around +90° and −90°, respectively. Orthogonal CP is thus produced by this arrangement at both the operating frequency bands. The connection between the complex amplitude of the incoming and the outgoing field may be found using the transmission matrix T in the following ways [36]:

$$\begin{bmatrix} E_x^t \\ E_y^t \end{bmatrix} = \mathrm{T} \begin{bmatrix} F_x^i \\ E_y^i \end{bmatrix} \tag{1.1}$$

$$= \begin{bmatrix} t_{xx} & t_{xy} \\ t_{yx} & t_{yy} \end{bmatrix} \begin{bmatrix} E_x^i \\ E_y^i \end{bmatrix}$$

where, $t_{ij} = E_i^t / E_j^i$, the polarized components of the transmitted and incident fields are denoted by the first and second subscripts, i and j, respectively. For

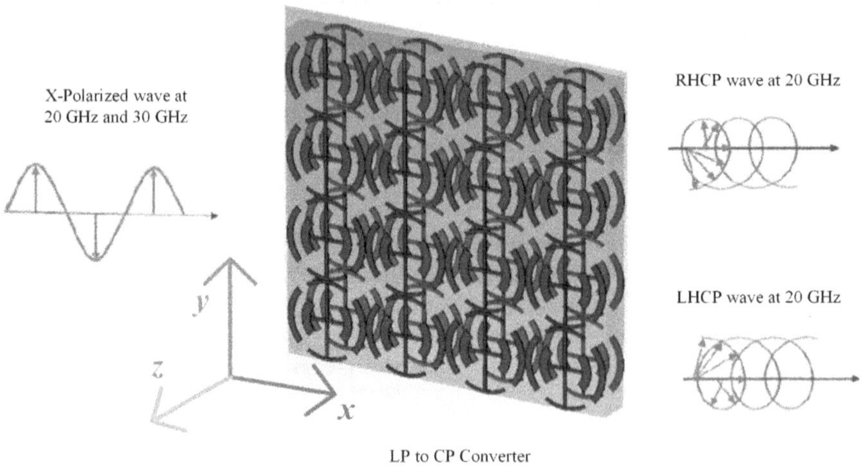

Figure 1.1 Topology of the suggested polarization converter.

each given transmitted wave, it can be composed of two equal-amplitude orthogonal components. Therefore,

$$\vec{E}^t = \vec{E}_x^t + \vec{E}_y^t = E_0^t\left(T_x\hat{x} + T_y\hat{y}\right)e^{-jkz} \tag{1.2}$$

where the linear transmission matrix is represented by $\begin{bmatrix} T_x \\ T_y \end{bmatrix} = \begin{bmatrix} |T_x|e^{j\varphi_x} \\ |T_y|e^{j\varphi_y} \end{bmatrix}$ and the amplitude by E_0^t. Consequently, at the converter's output, two orthogonal components undergo a phase shift of $\Delta\varphi = \varphi_x - \varphi_y$. As a result, the converter changes an incoming LP field into an outgoing field with CP if $|T_x| = |T_y|$ and $\Delta\varphi = \pm 90°$. This requirement needs to be met for a dual-band converter at both bands.

The expression for the transmission coefficient ($T.C.$) is:

$$T.C. = E^t / E^i \tag{1.3}$$

One way to characterize the transmitted wave with CP is as follows [30]:

$$\begin{bmatrix} E_{RHCP}^t \\ E_{LHCP}^t \end{bmatrix} = \begin{bmatrix} E_{RHCP}^t + jE_y^t \\ E_{LHCP}^t - jE_y^t \end{bmatrix} = (T.C.)\text{Circular}\begin{bmatrix} E_x^i \\ E_y^i \end{bmatrix}$$

$$(T.C.)\text{Circular} = \begin{bmatrix} T_{RHCPx} & T_{RHCPy} \\ T_{LHCPx} & T_{LHCPy} \end{bmatrix}$$

$$= 1/\sqrt{2}\begin{bmatrix} T_{xx} + jT_{yy} & T_{xy} + jT_{yy} \\ T_{xx} - jT_{yx} & T_{xy} - jT_{yy} \end{bmatrix} \tag{1.4}$$

This equation illustrates how the polarization converter may convert an LP wave into a CP wave where $1/\sqrt{2}$ in Equation (1.4) is caused by power normalization. Due to the insignificant cross-transmission coefficients of the suggested polarization converter ($T_{xy} \approx T_{xy} \approx 0$), we have

$$E^t_{RHCP} = 1/\sqrt{2}\left[(T_{xx})E^i_x + \left(jT_{yy}\right)E^i_y\right] \tag{1.5}$$

$$E^t_{LHCP} = 1/\sqrt{2}\left[(T_{xx})E^i_x - \left(jT_{yy}\right)E^i_y\right] \tag{1.6}$$

1.2.2 Design of unit cell

The proposed converter's fundamental unit, illustrated in Figure 1.2, consists of a solitary dielectric substrate layer, measuring 3.90 × 3.72 mm^2. With a thickness of 60 mil, dielectric constant 2.2, and tan δ of 0.0009, Diclad 880 is the selected substrate material. Transmission is enhanced by metallic split rings using patches printed on the two substrate sides. Copper, which is 0.035 mm thick, serves as the conductive substance

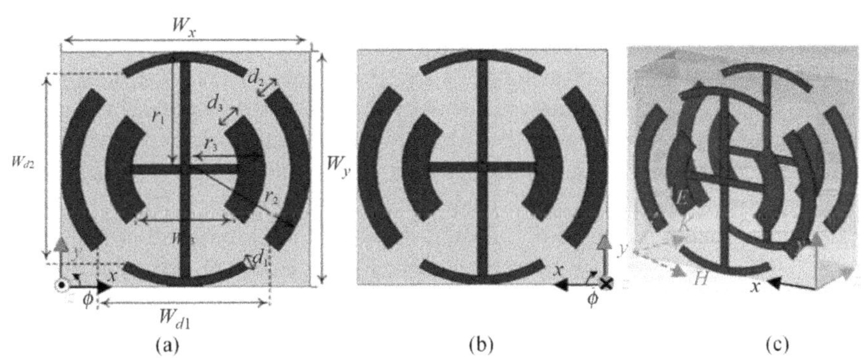

Figure 1.2 Unit-cell showing the (a) front, (b) backside, and perspective views of the proposed polarization converter. Dimensions: w_x = 3.90, w_y = 3.72, r_1 = 1.86, d_1= 0.15, w_{d1} = 2.2, r_2 = 1.95, d_2 = 0.3, w_{d2} = 3, r_3 = 1.25, d_3 = 0.42, w_{d3} = 1.35 (millimeter).

for all metal parts. By suitably altering the geometrical parameters, the converter's performance may be customized. Two orthogonal linearly polarized incoming waves cause the cell to react differently because of the asymmetry along the x- and y-axes. For the cell to act as an linear-to-circular (LTC) polarization converter, the two perpendicularly polarized fields are required to be transmitted with the same strength and a phase separation of 90°. The AR is applied to quantify the extent of CP and to graphically express CP waves. This is computed analytically using Equation (1.7) and takes into account the phase difference and transmission coefficient magnitude [37].

$$AR = 10\log_{10}\frac{\left(T_{xx}\cos\tau + T_{yy}\cos\Delta\varphi\,\sin\tau\right)^2 + T_{yy}^2\sin^2\Delta\varphi\,\sin^2\tau}{\left(T_{xx}\sin\tau + T_{yy}\cos\Delta\varphi\,\cos\tau\right)^2 - T_{yy}^2 sin^2\Delta\varphi\,\cos^2\tau} \tag{1.7}$$

where $\tau = 1/2\ \arctan\left((2T_{xx}\ T_{yy}\cos\Delta\varphi)/(T_{xx}^2 - T_{yy}^2)\right)$.

1.2.3 Behavior of the polarization converter with dual-band characteristics in the event of *a* = 45° polarized wave

Full-wave simulations are executed by means of Computer Simulation Technology (CST) Microwave Studio 2019 to properly examine and evaluate the performance of the dual-band LTC converter. Boundary conditions are specified as unit cell along the x- and y-directions, and open along the z-direction, to imitate an endless array of the structure. The behavior of the unit cell changes for two orthogonal linearly polarized waves due to non-uniform placement of components along the x- and y-axes. As shown in Figure 1.2, the cell is activated through incidence that is polarized along ϕ = 45°. The perpendicular fields of the transmission coefficients must be almost identical and in phase quadrature for the cell to operate as an LTC polarization converter. Figure 1.3(a) shows the magnitude of the transmission coefficients for the two polarizations. The transmission strength is still in excess of –0.5 dB for both frequency bands. Figure 1.3(b) shows the phase difference achieved for the dual-band LTC polarization converter. It is 78.2–107.45° in the lower frequencies ranging from 20.10 to 21.75 GHz and –75.72–95.66° in the higher frequencies from 29.12 to 31.4 GHz. It is evident that there are two nulls for T_{xx} in the transmission response. The structure's split rings act as coupled resonators for x-polarized waves and as nonresonant ones for y-polarization, resulting in the opposing polarizations of the transmitted CP waves in each of the bandwidths. This causes a 180° phase shift to occur in the transmission of T_{xx}.

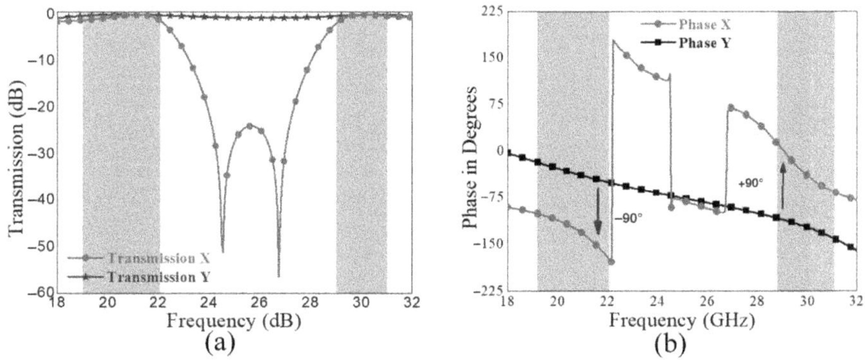

Figure 1.3 Simulated (a) transmission properties and (b) phase of the proposed polarization converter [30].

1.3 THE PROPOSED MIMO ANTENNA'S DESIGN AND CP MECHANISM

1.3.1 CP MIMO design with four ports

The dipole antenna possesses F-like structure displayed in Figure 1.4(a) and makes up the radiating element of the proposed MIMO antenna. Dual-band functionality is made possible by connecting two dipoles of varying dimensions to create this dipole. The striplines positioned on either side of the substrate layer feed the planar half-dipoles, creating two F-like dipoles in succession that are split apart by a Diclad 880 dielectric medium. The permittivity of the Diclad 880 substance is 2.2, and its tan δ of 0.0009. The radiating element investigation shows that dipole lengths L_1 and L_2 are the main factors influencing the antenna's operating frequency. The longer and shorter dipoles control the resonances at lower and upper frequencies, respectively. Figure 1.4(b) illustrates the concept of this CP antenna. A pair of single element dipole antennas arranged in parallel and spaced L_5 ($1.27\lambda_o$ at the antenna's lowest operational frequency) apart make up this MIMO antenna.

The dual-band transmissive LTC converter unit cell, as shown in Figure 1.4(c) and explained in Section 1.2, is utilized to change the LP wave that the dipole antenna emits into a CP wave. The converter's unit cell has a thickness of H_1, and it is developed by employing the dipole antenna's like substrate. On both sides of the substrate, split rings and metallic patches are printed which improve transmission. Additionally, in the antenna's endfire direction, a group of 7 × 7 units from LTC converter's unit cell is located distantly H_2 ($\lambda_o/4$) from the dipole. In order that the incidence field to contact

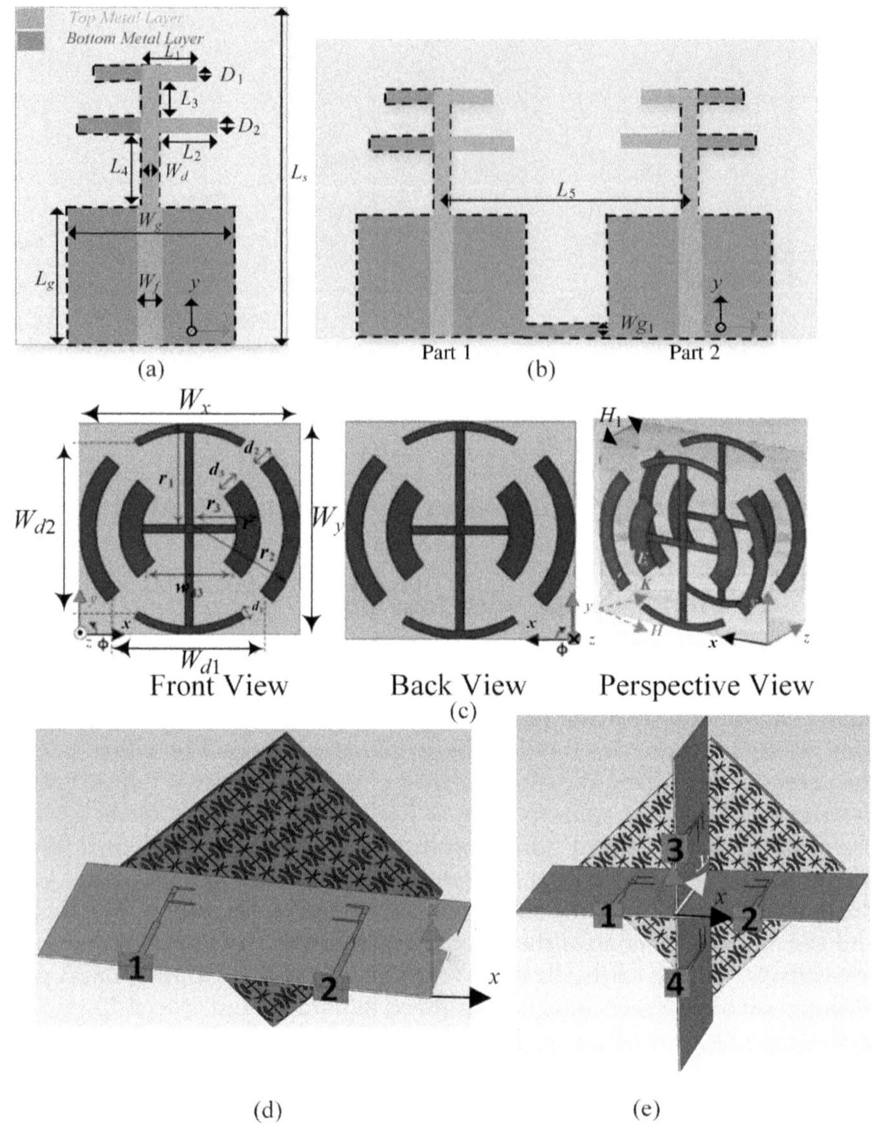

Figure 1.4 Schematic of (a) single element of MIMO, (b) two-port MIMO, and (c) unit cell of the double-band polarization converter [29]. Perspective view of the CP MIMO design with: (d) two-ports and (e) four-ports. D_1 = 0.5, L_1 = 2.05, D_2 = 0.4, L_2 = 3.13, W_d = 0.35, L_3 = 3.75, W_f = 0.75, W_g = 11, L_4 = 5.5, L_5 = 19.60, W_{g1} = 0.5, L_s = 23.85, L_g = 11, H_1 = 1.52 (millimeter).

the LTC at 45°, it is revolved in relation to the z-axis. Figure 1.4(d) shows the initial design of a two-port CP MIMO antenna. By adding other similar MIMO structure orthogonal to it, this design is then extended to a four-port configuration. With this configuration, as shown in Figure 1.4(e), all four MIMO components are positioned orthogonally to one another and share a common ground plane.

A LTC polarization converter is formed by the 7 × 7 group of units shown in Figure 1.4(c), which is oriented in the antenna's radiating direction. With a phase difference of ±90°, the dipole antenna's LP wave is split into two orthogonal components of identical amplitudes by the converter. A CP wave within the operating frequency range is produced as an outcome of this technique. For both x- and y-polarizations, the transmission magnitudes are almost equal in band of frequencies from 19.0 to 22.0 GHz and 29.0 to 31.0 GHz, as depicted in Figure 1.3(a). Furthermore, as shown in Figure 1.3(b), the phase difference between them is −90° in the lower frequency range and +90° in the upper frequency band, respectively. The observed features do, in fact, satisfy the requirements for the conversion to CP, as given by $|T_x| = |T_y|$ and phase difference between them as ±90°, where transmitted parameters along the x- and y-directions, respectively are denoted by T_x and T_y. Remarkably, the orthogonal CPs may be switched around simply changing the incident wave's polarization.

1.4 VALIDATION OF FOUR-PORT CIRCULARLY POLARIZED MIMO ANTENNA: NUMERICAL ANALYSIS AND EXPERIMENTAL OUTCOMES

The CP MIMO antenna under consideration is modeled and simulated utilizing CST Microwave Studio. Figure 1.5(a) illustrates the perspective view of the proposed MIMO antenna. Subsequently, a prototype of it is fabricated through the photolithographic procedure, as depicted in Figure 1.5(b).

Diclad 880 dielectric material is used to manufacture the MIMO structure and a group of 7 × 7 units of the cells of the polarization converter. Figure 1.5(c) presents the measurements obtained in the anechoic chamber. The thicknesses of these elements are 10 mil and 60 mil, correspondingly. Figure 1.6(a)–(d) shows the simulated and tested reflection characteristics of the two-port and four-port CP MIMO antennas, respectively. Two resonant frequencies are clearly attained, with a center at 20.0 GHz and a second at 30.25 GHz. Due to the two dipole's differing lengths in the cascaded design, there are two resonance frequencies.

The distribution of surface current over the dipole, as shown in Figure 1.6(c), further supports this conclusion. At 20 and 30 GHz, respectively, the current is mostly focused across the lengths L_2 and L_1. Consequently, it can be inferred that the lengths L_2 and L_1 provide the two resonant frequencies.

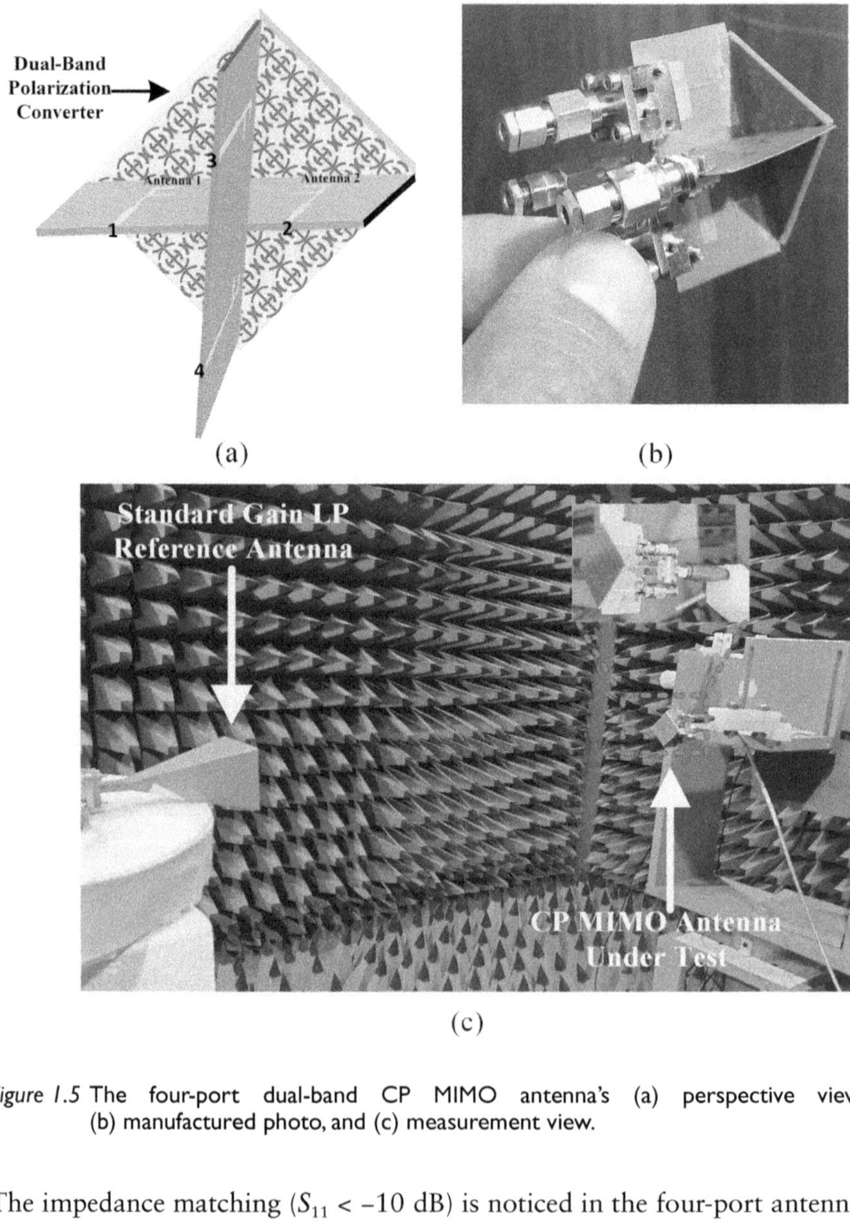

Figure 1.5 The four-port dual-band CP MIMO antenna's (a) perspective view, (b) manufactured photo, and (c) measurement view.

The impedance matching ($S_{11} < -10$ dB) is noticed in the four-port antenna within two frequency bands. In the lower frequency band, impedance matching is achieved from 18.62 to 20.40 GHz, resulting in a fractional impedance bandwidth of 9.12%. Similarly, impedance matching is noted in the higher frequency range of 27.15–30.35 GHz, which corresponds to a fractional impedance bandwidth of 11.13%. The two main Sat-Com bands

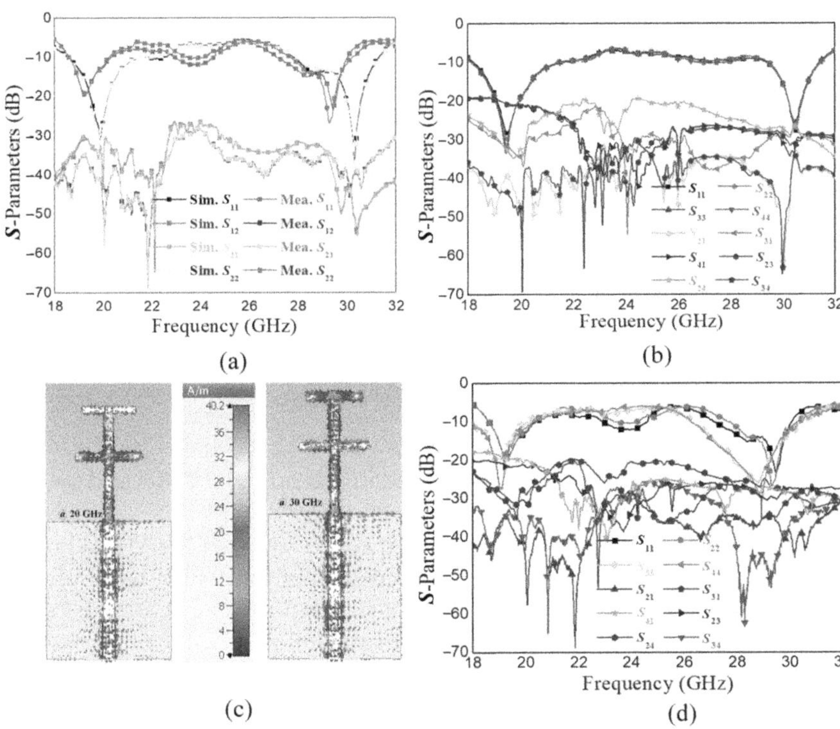

Figure 1.6 (a) Simulated and tested S-parameters of two-port circularly polarized MIMO antenna. Four-port circularly polarized MIMO antenna S-parameters (b) simulated and (d) measured. (c) A single element of the dual-band CP MIMO antenna's distribution of surface current at 20 and 30 GHz.

that are allocated to fixed satellite services—upstream (29.5–30.0 GHz) and downstream (19.7–20.2 GHz) are covered by the antenna. As shown in Figure 1.6(c), the observed isolation between every port is over and above -19 dB in 18.62–20.40 GHz and 27.15–30.35 operating bands.

The CP functionality of an antenna as indicated by the AR is shown in Figure 1.7(a) and (b). Both the frequency ranges 19.65–22.05 GHz and 29.25–30.35 GHz show good agreement between the simulated and observed ARs. As a result, observed AR bandwidths of 11.51% (19.65–22.05 GHz) and 3.69% (29.25–30.35 GHz) are achieved by the developed double-band, four-port CP MIMO antenna. In the frequencies of 19.65–20.40 GHz and 29.25–30.35 GHz, the antenna yields common AR and impedance bandwidths, respectively, as 3.74% and 3.69%. Figure 1.7(a) and (b) show the gain as a function of frequency for the proposed MIMO antenna in the frequency bands 18.0–22.0 GHz and 28.0–32.0 GHz,

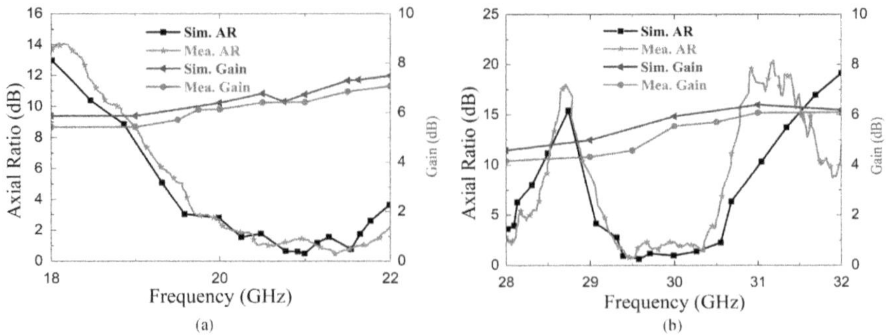

Figure 1.7 Gain and AR of a four-port, dual-band, circularly polarized MIMO antenna at (a) lower frequency band and (b) upper frequency band.

respectively. The observed gain, measured in relation to $\theta = 90°$ and $\varphi = 90°$, is 5.75 dBic at 20 GHz and 4.77 dBic at 30 GHz.

In an MIMO arrangement, the isolation effectiveness of the MIMO unit is measured by the cross-correlation (CC) coefficient (ρ), which has an immediate effect on the maximum channel capacity that may be achieved. Perfect isolation between antenna elements is indicated by a CC value of zero, which is ideal for an MIMO antenna. However, as it indicates comparatively low correlation between antenna elements, a CC value of less than 0.5 is often accepted for MIMO systems. Equation (1.8) [38] is used to evaluate the CC of MIMO making use of far-field antenna properties. In this equation, Ω stands for the solid angle, E_{porti} for the antenna's far-field when port i is stimulated, and E_{portj} for the antenna's far-field when port j is fed.

$$\rho_{porti,portj} = \frac{\frac{1}{2Zo}\iint E_{porti} \cdot E^{*}_{portj} d\Omega}{\sqrt{P_{rad,porti}}\sqrt{P_{rad,portj}}} \tag{1.8}$$

As shown in Figure 1.8(a) and (b), the CC between each antenna in the specified four-port CP MIMO scheme stays inside allowable bounds in both the operating frequency bands. Diversity gain (DG) is an additional MIMO system metric that illustrates how diversity design affects radiated power. The correlation matrix R and Frobenius norm is used to determine the DG of the MIMO using (1.9) [39]

$$G = \frac{tr(R)^2}{\|R\|_{F_r}} \tag{1.9}$$

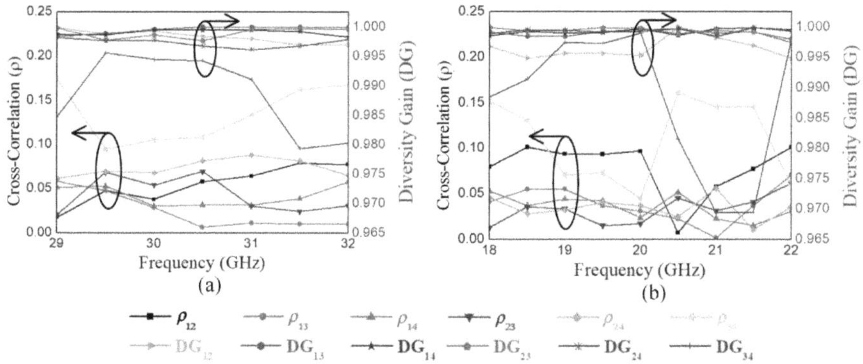

Figure 1.8 Double-band, four-port CP MIMO antenna's CC and DG at (a) the lower and (b) upper frequency bands.

where $\|R\|_{F_r}$ is the Frobenius norm defined as $\|R\|_{F_r} := \sum_{m,n} |\rho_{mn}|^2$ and *tr* is the matrix trace. The diagonal components of matrix R^2 added together make up the matrix trace. Generally speaking, the DG of an MIMO antenna falls between 0 and 1 [39]. Figure 1.8(a) and (b) depict the DG fluctuation, respectively, at the operating frequency bands of the proposed MIMO antenna. The MIMO structure's simulated DG is greater than 0.99 in both frequency ranges.

Furthermore, Figure 1.9 depicts the gain plot and radiation pattern of the individual unit of the suggested four-port MIMO antenna in the *xy* plane. As shown in Figure 1.9(b), the gain for the individual unit in both the lower and higher frequency bands is 5.92 dBic and 6.10 dBic, respectively, measured along θ = 90° and φ = 90°. As illustrated in Figure 1.9(c) and (d), there exists approximately a 20 dB difference between the right-hand circular polarization (RHCP) and left-hand circular polarization (LHCP) radiation patterns at both operating frequencies, indicating excellent CP performance. By providing a −90° and +90° phase difference between the two transmission components in the lower and upper frequency bands, respectively, the polarization converter allows for the achievement of this orthogonal CP. Figure 1.10(a) and (b) at 20 GHz and Figure 1.10(c) and (d) at 30 GHz, respectively, show the computed and observed far-field radiation patterns in the *xy* plane of the four-port CP MIMO antenna system by stimulating ports one and three.

There is a noticeable similarity between the measured and simulated findings. It is apparent that by stimulating the orthogonal ports, the CP characteristic potentially be inverted at both the operating frequency bands. Moreover, because of the MIMO components' symmetry, a similar radiation

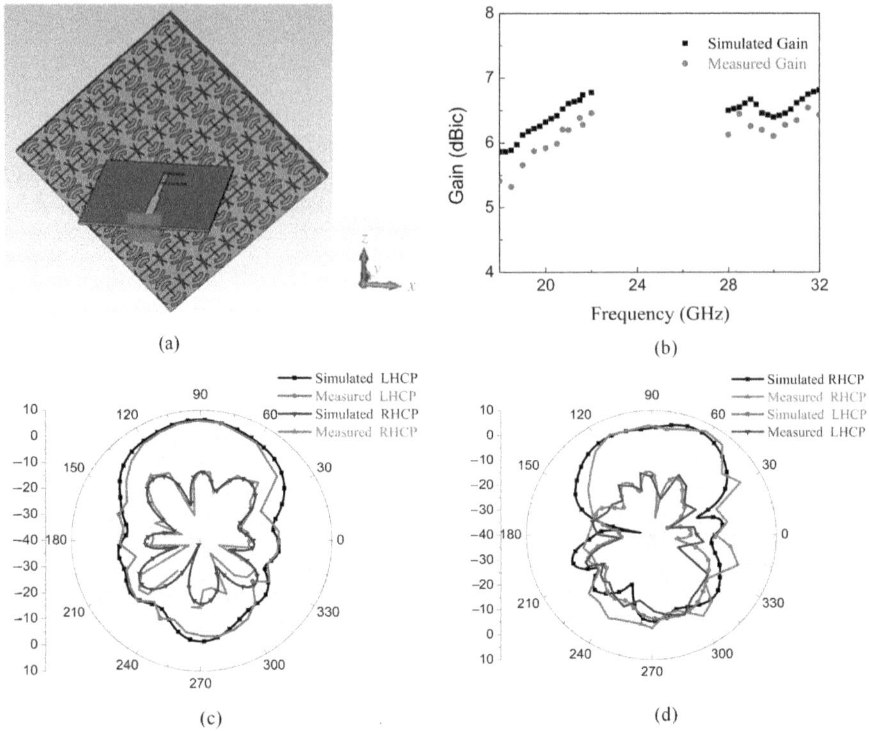

Figure 1.9 (a) The single element of the suggested CP MIMO antenna; (b) the simulated and observed gain and the radiation pattern in the *xy* plane at (c) 20 and (d) 30 GHz.

pattern for both frequency bands is achieved at ports two and four. Table 1.1 compares the suggested CP MIMO antenna's performance to that of other CP MIMO antennas [15]–[28]. Orthogonal CP is present in both the lower and upper frequency bands of the proposed dual-band CP MIMO antenna. It is also possible to obtain polarization and spatial diversity because of the design of the proposed CP MIMO antenna.

1.5 CONCLUSION

This study presents a unique four-port, double-band CP MIMO antenna that integrates a 7 × 7 group of units of the double-band LTC polarization converter with two consecutive F-like dipoles. The LTC polarization converter is used by this MIMO antenna to produce orthogonal CP in the frequency bands 29.25–30.35 GHz and 19.65–22.05 GHz. The proposed CP MIMO antenna prototype is fabricated and measured; the results show that the measured CP overlapping bandwidths in frequency ranges of

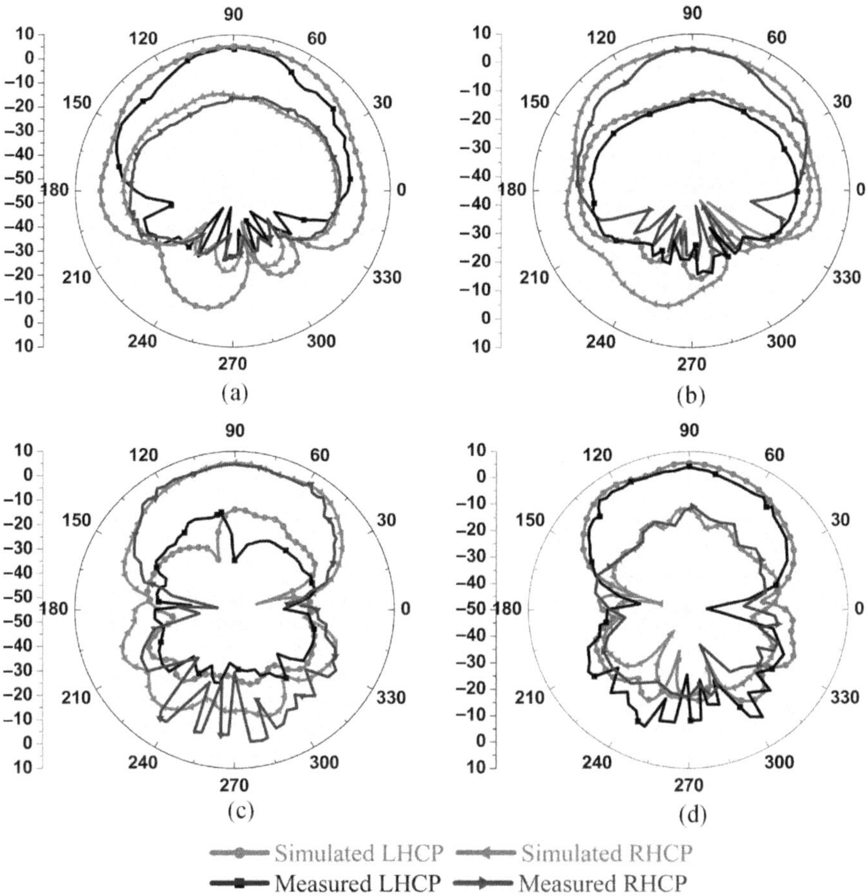

Figure 1.10 Double-band CP MIMO antenna radiation pattern of in the *xy* plane at 20 GHz for (a) port 1 and (b) port 3, and 30 GHz for (c) port 1 and (d) port 3, respectively.

18.62–20.40 GHz and 29.25–30.35 GHz are 3.69% and 3.74%, respectively. The bandwidths that these surpass are needed to meet the uplink and downlink frequency requirements for Ka-band fixed satellite services. At both operating bands, the suggested antenna has exceptional MIMO characteristics, such as isolation of better than –19 dB, CC lower than 0.19, and DG better than 0.97. Furthermore, the proposed CP MIMO antenna achieves an observed gain respectively, as 5.75 dBic and 4.77 dBic at 20 GHz and 30 GHz. All things considered, presented MIMO structure has potential for use in Sat-Com on moving vehicles. Furthermore, the presented antenna is utilizable as feed source to transmitarray/reflectarray operating at the working frequency bands of the antenna.

Table 1.1 Evaluation of the proposed dual-band CP MIMO antenna's performance in comparison to existing MIMO antennas found in published works

Reference	*Frequency (GHZ)*	*CP Method*	*Polarization*	*Size (λ_o^2)*	*CP Bandwidth (%)*	*Number of Ports*	*Gain (dB)*	*Radiation Efficiency (%)*
[15]	6	Dual Port Feed Network	LHCP	0.8 × 1.3	19.9	2	13.9	—
[16]	3.7, 4.9	Unequal Feedline Mechanism	LHCP, LHCP	—	5.4, 4.08	2	3.75, 5	—
[17]	23.8, 27.93	Aperture Excitation	LHCP, Linear	0.456 × 0.456	2.1	2	6.53	—
[18]	27.3	Corner Truncation	—	1 × 1	16.8	4	9.5–11	92
[19]	2.33	Corner Truncation	LHCP	0.67 × 0.67	43.33	4	1.96– 2.36	—
[20]	2.45	Inductor Loaded Metal Strip	LHCP	0.5 × 0.5	4.08	2	—	40
[21]	2.54, 3.52	Aperture Excitation	RHCP, RHCP	0.67 × 0.42	7.89,7.09	2	5.9, 6	95
[22]	7.88	Element Geometry	RHCP	2.1 × 2.1	4.55	2	3.8	—
[23]	3.6	Open Slot in the Ground Plane	LHCP, RHCP	0.72 × 0.72	—	4	4.5	—
[24]	3.9	—	Linear	1.04 × 1.04	—	4	4–6	93
[25]	3.25	—	Linear	0.52 × 1.05	—	4	3–4	80–92
[26]	30	Aperture Excitation	LHCP	2.7 × 2.7	6	4	—	—
[27]	2.375	Using Four Microstrip Arcs	RHCP	0.55 × 0.55	7.71	2	0.87–1.48	88
[28]	28	Corner Truncation	RHCP	1.58 × 1.58	17	2	14.1	—
This Work	19.51, 28.67	Linear to Circular Polarization Converter	LHCP, RHCP	1.55 × 2.92	11.51, 3.69	4	5.75, 4.77	>97

REFERENCES

[1] C.A. Balanis, Antenna Theory Analysis and Design (4th ed.). John Wiley & Sons, pp. 70–75, 811–876, 2005.
[2] C. Dietlein, A. Luukanen, Z. Popovi, and E. Grossman, "A W-band polarization converter and isolator," IEEE Trans. Antennas Propag., vol. 55, no. 6, pp. 1804–1809, Jun. 2007.
[3] A. Gerald, and J.M. Burch, Introduction to Matrix Methods in Optics (1st ed.). John Wiley & Sons, 1975. ISBN 0471296856.
[4] D.S. Kliger, J.W. Lewis, and C.E. Randall, "Introduction to the Jones Calculus, Mueller Calculus, and Poincare Sphere," in Polarized Light in Optics and Spectroscopy, pp. 59–101, 1990, Elsevier, https://doi.org/10.1016/B978-0-08-057104-1.50007-7
[5] L. Htein, D.S. Gunawardena, W.-H. Chung, H.-Y. Au, and H.-Y. Tam, Accelerometer employing a side-hole fiber in a Sagnac interferometer, J. Lightw. Technol., vol. 39, no. 10, pp. 3303–3311, May, 2021, doi: 10.1109/JLT.2021.3056474.
[6] L.J. Aplet and J.W. Carson, "A Faraday effect optical isolator," Appl. Opt., vol. 3, no. 4, pp. 544–545, 1964.
[7] E. Prati, "Propagation in gyroelectromagnetic guiding systems," J. Electr. Wav. Appl., vol. 17, no. 8, pp. 11771196, 2003.
[8] J.B. Pendry,, A.J. Holden, , D.J. Robbins, and W.J. Stewart, "Magnetism from conductors and enhanced nonlinear phenomena," IEEE Trans. Microw. Theory Techn., vol. 47 no. 11, pp. 2075–2084, 1999.
[9] R.A. Shelby, D.R. Smith, and S. Schultz, "Experimental verification of a negative index of refraction," Science, vol. 6, pp. 77 – 91, 2001. doi: 10.1126/science.1058847. PMID: 11292865.
[10] C. Jansen, I.A. Al-Naib, N. Born, and M. Koch, "Terahertz metasurfaces with high Q- factors," Appl. Phys. Lett., vol. 98, no. 5, 2011.
[11] W.-M. Li, B. Liu, and H.-W. Zhao, "The U-shaped structure in dual-band circularly polarized slot antenna design," IEEE Antennas Wireless Propag. Lett., vol. 13, pp. 447–450, 2014.
[12] K. Saurav, D. Sarkar, and K.V. Srivastava, "Dual-band circularly polarized cavity-backed crossed-dipole antennas," IEEE Antennas Wireless Propag. Lett., vol. 14, pp. 52–55, 2015.
[13] S. Mener, R. Gillard, and L. Roy, "A dual-band dual-circular-polarization antenna for Ka-band satellite communications," IEEE Antennas Wireless Propag. Lett., vol. 16, pp. 274–277, 2017.
[14] C.M. Su and K.L. Wong, "A dual-band GPS microstrip antenna," Microw. Opt. Techn. Lett., vol. 33, no. 4, pp. 238–240, 2002.
[15] S. Karamzadeh and M. Kartal, "Circularly polarised MIMO tapered slot antenna array for C-band application," Electron. Lett., vol. 51, no. 18, pp. 1394–1396, Sep. 2015.
[16] S. Chakraborty, M.A. Rahman, M.A. Hossain, E. Nishiyama, and I. Toyoda, "Design of a 5G compatible two-element dual-band circularly polarized MIMO antenna," in Proc. IEEE Asia–Pacific Microw. Conf. (APMC), Singapore, pp. 923–925, IEEE, Dec. 2019.

[17] A. Gaya, M.H. Jamaluddin, I. Ali, and H. Mohamad, "Dual band circularly polarized rectangular dielectric resonator antenna for millimeter wave 5G applications," In Proc. IEEE Asia–Pacific Conf. Appl. Electromagn. (APACE), Melacca, Malaysia, pp. 1–5, IEEE, Nov. 2019.

[18] N. Hussain, M.J. Jeong, A. Abbas, and N. Kim, "Metasurface-based single-layer wideband circularly polarized MIMO antenna for 5G millimeter-wave systems," IEEE Access, vol. 8, pp. 130293–130304, 2020.

[19] A. Iqbal, A. Smida, A.J. Alazemi, M.I. Waly, N.K. Mallat, and S. Kim, "Wideband circularly polarized MIMO antenna for high data wearable biotelemetric devices," IEEE Access, vol. 8, pp. 17935–17944, 2020.

[20] L. Qu, H. Piao, Y. Qu, H.-H. Kim, and H. Kim, "Circularly polarised MIMO ground radiation antennas for wearable devices," Electron. Lett., vol. 54, no. 4, pp. 189–190, Feb. 2018.

[21] G. Bharti, D. Kumar, A.K. Gautam, and A. Sharma, "Two-port dual-band circularly polarized dielectric resonator-based MIMO antenna with polarization diversity," Electromagnetics, vol. 40, no. 7, pp. 463–478, Oct. 2020.

[22] G. Varshney, R. Singh, V.S. Pandey, and R.S. Yaduvanshi, "Circularly polarized two-port MIMO dielectric resonator antenna," Prog. Electromagn. Res. M, vol. 91, pp. 19–28, 2020.

[23] S. Saxena, B.K. Kanaujia, S. Dwari, S. Kumar, H.C. Choi, and K.W. Kim, "Planar four-port dual circularly-polarized MIMO antenna for sub-6 GHz band," IEEE Access, vol. 8, pp. 90779–90791, 2020.

[24] G. Varshney, S. Gotra, S. Chaturvedi, V.S. Pandey, and R.S. Yaduvanshi, "Compact four-port MIMO dielectric resonator antenna with pattern diversity," IET Microw. Antennas Propag., vol. 13, no. 12, pp. 2193–2198, 2019.

[25] H.T. Chattha, "4-port 2-element MIMO antenna for 5G portable applications," IEEE Access, vol. 7, pp. 96516–96520, 2019.

[26] M. Akbari, H.A. Ghalyon, M. Farahani, A.R. Sebak, and T.A. Denidni, "Spatially decoupling of CP antennas based on FSS for 30-GHz MIMO systems," IEEE Access, vol. 5, pp. 6527–6537, 2017.

[27] W.W. Li, K.W. Leung, and N. Yang, "Omnidirectional dielectric resonator antenna with a planar feed for circular polarization diversity design," IEEE Trans. Antennas Propag., vol. 66, no. 3, pp. 1189–1197, Mar. 2018.

[28] N. Hussain, M.J. Jeong, J. Park, and N. Kim, "A broadband circularly polarized Fabry–Perot resonant antenna using a single-layered PRS for 5G MIMO applications," IEEE Access, vol. 7, pp. 42897–42907, 2019.

[29] M.A. Sofi, K. Saurav, and S.K. Koul, "Four-Port Orthogonal Circularly Polarized Dual-Band MIMO Antenna With Polarization and Spatial Diversity Using a Dual-Band Linear-to-Circular Polarization Converter," IEEE Trans. Antennas Propag., vol. 70, no. 9, pp. 8554–8559, Sept. 2022.

[30] M.A. Sofi, K. Saurav, and S.K. Koul, "Frequency-selective surface-based compact single substrate layer dual-band transmission-type linear-to-circular polarization converter," IEEE Trans. Microw. Theory Techn., vol. 68, no. 10, pp. 4138–4149, Oct. 2020.

[31] Y.X. Zhang, Y.C. Jiao, and L. Zhang, "Wideband circularly polarized array antennas with sequential-rotation polarization grid and simplified full-SIW feeding networks," IEEE Trans. Antennas Propag., vol. 68, no. 8, pp. 6088–6097, Aug. 2020.
[32] J. Wu, Y.J. Cheng, and Y. Fan, "Millimeter-wave wideband high- efficiency circularly polarized planar array antenna," IEEE Trans. Antennas Propag., vol. 64, no. 2, pp. 535–542, Feb. 2016.
[33] C. Ma, Z.H. Ma, and X. Zhang, "Millimeter-wave circularly polarized array antenna using substrate-integrated gap waveguide sequentially rotating phase feed," IEEE Antennas Wireless Propag. Lett., vol. 18, no. 6, pp. 1124–1128, Jun. 2019.
[34] T.L. Zhang, L. Chen, S.M. Moghaddam, A.U. Zaman, and J. Yang, "Millimeter-wave ultrawideband circularly polarized planar array antenna using bold-C spiral elements with concept of tightly coupled array," IEEE Trans. Antennas Propag., vol. 69, no. 4, pp. 2013–2022, Apr. 2021.
[35] H. Al-Saedi et al., "A low-cost Ka-band circularly polarized passive phased-array antenna for mobile satellite applications," IEEE Trans. Antennas Propag., vol. 67, no. 1, pp. 221–231, Jan. 2019.
[36] S. Yan and G.A.E. Vandenbosch, "Compact circular polarizer based on chiral twisted double split-ring resonator," Appl. Phys. Lett., vol. 102, no. 10, p. 103503, Mar. 2013.
[37] W. Li et al., "A reconfigurable polarization converter using active metasurface and its application in horn antenna," IEEE Trans. Antennas Propag., vol. 64, no. 12, pp. 5281–5290, Dec. 2016.
[38] K. Saurav, N.K. Mallat, and Y.M.M. Antar, "A three-port polarization and pattern diversity ring antenna," IEEE Antennas Wireless Propag. Lett., vol. 17, no. 7, pp. 1324–1328, Jul. 2018.
[39] S.M. Mikki and Y.M.M. Antar, "On cross correlation in antenna arrays 374 with applications to spatial diversity and MIMO systems," IEEE Trans. Antennas Propag., vol. 63, no. 4, pp. 1798–1810, Apr. 2015.

Chapter 2

Design and optimization of metasurface antennas in wireless communication

Diptiranjan Samantaray and Somak Bhattacharyya

2.1 INTRODUCTION TO METASURFACE

This section would introduce the concept of metasurfaces (MS), which are artificially engineered surfaces with subwavelength structures that manipulate electromagnetic (EM) waves in unconventional ways. MS take part in a major role due to the incorporation of cutting-edge methodologies, concepts, information, and design styles for a variety of microwave devices [1–5]. They are composed of a periodic group of conductive elements, which include metal rings, rods or various geometric shapes forming an EM medium at the same time, and their resonance effect is determined by the geometry of unit cells [6–14]. The MS elements are designed to achieve phase control across a wide frequency band, enabling beamforming and steering capabilities [15–25].

2.2 METASURFACE ANTENNA DESIGN TECHNIQUES

Metasurface antennas (MSA) are a class of antennas that utilize engineered structures, known as MS, to manipulate EM waves in desired ways [3, 7, 17, 18]. These MS typically consist of subwavelength structures arranged in periodic or quasi-periodic patterns that allow for control of the phase, amplitude, and polarization of EM waves [7, 8]. MSA offer advantages such as compact size, low profile, and the ability to tailor radiation patterns and polarization responses, making them suitable for various applications [7, 8].

2.2.1 Design considerations for metasurface antenna

Miniaturization: MSA enable the miniaturization of antenna systems while maintaining or even enhancing performance characteristics. This is particularly crucial for applications where space constraints or form factor limitations are a concern [3, 7, 8].

Customizability: The ability to precisely control the properties of EM waves, such as phase, polarization, and propagation direction, allows for

DOI: 10.1201/9781003470854-2

the customization of antenna functionalities to meet specific requirements and application scenarios [3, 7, 8].

Bandwidth and Efficiency: MSA can exhibit wideband operation and high efficiency, addressing the demands of modern communication systems for robust and reliable performance across broad frequency ranges [8–12]. Wideband operation is crucial for practical MSA, especially for applications like 5G communication [13]. Design strategies such as multilayered MS or broadband unit cells are employed to achieve wideband performance. Maximizing antenna efficiency is essential to ensure reliable communication. Careful design optimization, material selection, and fabrication techniques are employed to minimize losses and enhance efficiency [26–30].

Integration: Practical MSA must be compatible with existing communication systems and devices. Considerations for integration with circuitry, feeding networks, and radomes are vital to ensure seamless integration into practical systems. MSA can be seamlessly integrated into various platforms, including consumer electronics, Internet of Things (IoT) devices, and aerospace systems, due to their planar and conformal nature [31–49].

2.3 METASURFACE SYNTHESIS AND CIRCUIT MODEL

Metasurfaces are gaining popularity due to their ease of manufacture and wide range of practical applications [50–52]. A propagating wave encounters discontinuity when an MS is present. The analysis of an MS can also be done by using a circuit model, where the MS is represented as a shunt impedance in the study [50–52]. The reactance X describes the inductive or capacitive behavior of the periodic surface, while the real part of the equivalent impedance R corresponds to the ohmic losses in the metal and dielectric layers. The MS reflection and transmission coefficients (S_{11} and S_{21}) can be determined using typical transmission line relations as provided in (9.1 and 9.2), respectively

$$S_{11} = \frac{AZ_0 + B - CZ_0^2 - DZ_0}{AZ_0 + B + CZ_0^2 + DZ_0} \tag{9.1}$$

$$S_{21} = \frac{2Z_0}{AZ_0 + B + CZ_0^2 + DZ_0} \tag{9.2}$$

where A, B, C, D represent the parameters of the $ABCD$ matrix of the model containing MS. The $ABCD$ matrix is mentioned in (9.3)

$$\begin{bmatrix} A & B \\ C & D \end{bmatrix} = [M_1] \cdot \ldots \cdot [M_n] \cdot [M_{\mathrm{MS}}] \cdot [M_{n+1}] \cdot \ldots \cdot [M_N] \tag{9.3}$$

Here, M are the $ABCD$ matrix of every layer, N is the generic number of the dielectrics, and n represents the nth dielectric substrate. Each layer for nth order element and for MS are represented in (9.4 and 9.5)

$$[M_n] = \begin{bmatrix} \cos(k_{zn} d_n) & jZ_n \sin(k_{zn} d_n) \\ j\dfrac{\sin(k_{zn} d_n)}{Z_n} & \cos(k_{zn} d_n) \end{bmatrix} \tag{9.4}$$

$$[M_{\mathrm{MS}}] = \begin{bmatrix} 1 & 0 \\ 1/Z_{\mathrm{MS}} & 1 \end{bmatrix} \tag{9.5}$$

where Z_{MS} is an approximate MS impedance computed from retrieved L and C parameters as mentioned in Table 2.1. $Z_n^{\mathrm{TE}} = \dfrac{(\omega\mu_r\mu_0)}{k_{zn}}$; $Z_n^{\mathrm{TM}} = \dfrac{k_{zn}}{(\omega\varepsilon_r\varepsilon_0)}$ are the characteristic impedance of the dielectric layer for transverse electric (TE) and transverse magnetic (TM) polarization, respectively, $k_{zn} = \sqrt{\epsilon_r k_0^2 - k_t^2}$ is the wavenumber and $k_t = k_0 \sin(v)$ is the transverse component of the wavenumber with v corresponding to the normal incidence angle of the incoming wave, and k_0 is the free space propagation constant. The quantities ϵ_0, ϵ_0, μ_0, and μ_r are the free space and relative dielectric permittivity and magnetic permeability, respectively.

A periodic unit cell can be built by an unlimited number of geometries. Certain geometries are more popular than others and are also straightforward to comprehend and manipulate. It is proved that the MS reactance expression obeys Foster's theorem, implying that it has the identical pole-zero analytic features as a passive LC network. The quantity of lumped elements is proportional to the number of resonances [50–52].

Various MS elements are depicted in Figures 2.1 and 2.2 along with their corresponding circuit representations. The impedance of basic nonresonant components like wire grids or patch arrays is predominantly inductive or capacitive. When a resonant element is employed instead of a purely capacitive or inductive one, a straightforward LC circuit can be utilized to match the frequency response of an MS [53–55]. In the case of capacitive MS, the circuit will be inductive in series, while an inductive MS will utilize a shunt

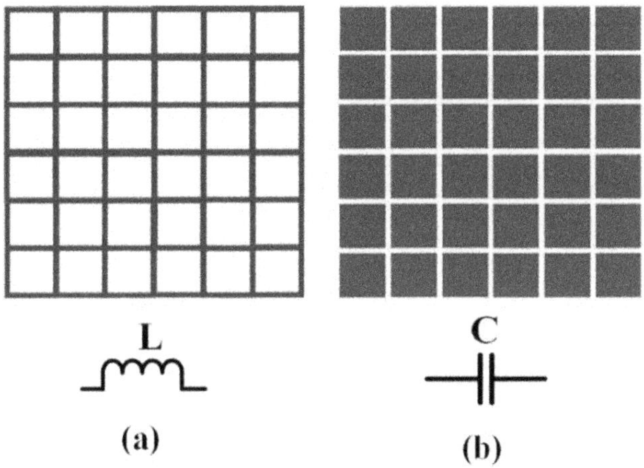

Figure 2.1 Metasurface elements and associated single element circuit models (a) inductive grid (b) capacitive patch.

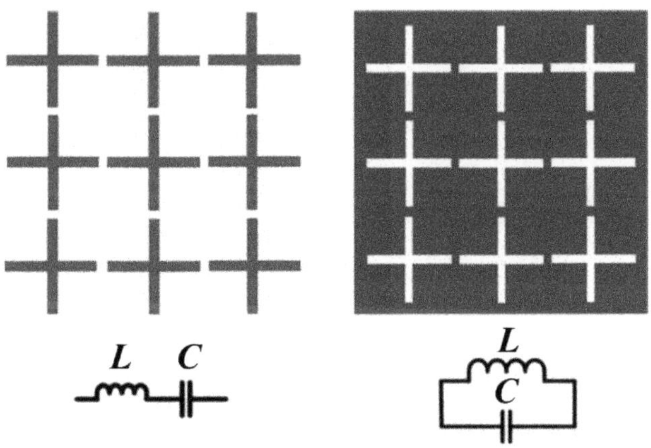

Figure 2.2 Metasurface elements and associated multiple element circuit models (a) capacitive cross and (b) inductive cross.

LC configuration. The MS impedance is entirely imaginary in the absence of losses and is characterized by two lumped parameters. Losses can be incorporated into an equivalent circuit by introducing a series resistance. Table 2.1 outlines the classification of the most common geometries based on resonant characteristics and equivalent circuits [50–52].

Table 2.1 Metasurface elements classification and circuit model topology

Element type	*Element shape*	*Equivalent circuit*
Nonresonant elements	Strip, patch, wire grid	C
Single-resonant	Loop, dipole, cross, tripole	(LC)
Double-resonant	Double cross, cross frame, double loop, Jerusalem cross, etc.	(LC)//(LC) or (LC)(L//C)
Multi-resonant	Concentric loop, fractal elements, quadrifilar spiral, meandered dipole	(LC)//(LC)//(LC)....
Other	Split ring, cut loop	-

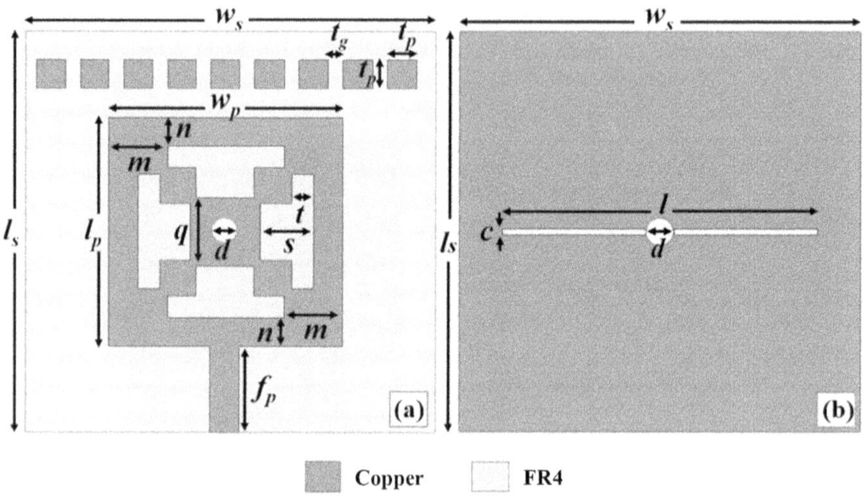

Figure 2.3 Geometry of the antenna structure with a top view (a) and rear view (b) of the conventional patch [11].

2.4 DESIGN OF THE PRACTICAL METASURFACE ANTENNA

In this section, Figure 2.3 shows a portable, directed, and highly effective MS-based fractal antenna in a stacked arrangement with the radiating patch and the MS. The patch uses dielectric vias and fractal geometry to produce the appropriate frequency range, which enables effective frequency modulation [11].

The shunt inductive effect dielectric via reduces the electric field around it. As a result, the field distribution is strongly perturbed that leads to tune-up of its resonant frequency, resulting in an increase in the electrical length of the patch and enhancing the radiation directivity. The reflection coefficient is also improved by the optimal position of the via on the middle line of the patch. The prototype utilized a defected ground concept along with the addition of two rectangular slots on the floor plane. This led to a decrease in surface wave radiation, consequently reducing cross-polarized radiation. Slots on the ground plane are a load introduced to the antenna, which can be used to increase the input impedance to the characteristic impedance of 50 ohms. This improves the input impedance matching and thus reflection characteristics, $|S_{11}|$. Slots are placed in such a way that they reduce the radiation so that the effect of the slot can be represented by a purely reactive impedance [56–58]. Depending on the antenna geometry and position of the slot, the input impedance will be influenced differently. In the prototype, a flawed ground concept was implemented, along with the incorporation of two rectangular slots into the floor plane. Consequently, surface wave radiation is diminished, thereby minimizing cross-polarized radiation.

Furthermore, an array of square patches consisting of 1 × 9 elements has been incorporated into the upper section of a fractal patch. This inclusion leads to a 7.26% increase in fractional bandwidth, reaching 11.28 GHz, and an optimum gain of 5.90 dBi at 10.78 GHz. Each individual square patch exhibits an inductive response, while the gap between two adjacent square patches demonstrates a capacitive response. The 1 × 9 array of square patches forms a cascaded LC circuit, with each LC circuit resonating at a specific frequency. When combined with the fractal patch, the cascaded LC network forms an equivalent circuit, effectively expanding the antenna's bandwidth by 7.26%. This antenna has been widely utilized as a conventional patch antenna in various device applications. The antenna gain is further improved by the addition of a 5 × 5 order mesh that acts as the superstrate for the patch. In this scenario, the resonant effect of the incident EM wave is controlled by the geometry of the mesh structure arranged in the superstrate manner. Both planar and nonplanar configurations of MS designs usually incorporate high-impedance surfaces, which effectively modulate the EM fields generated by the patch resonator [59–63].

2.4.1 Design of the multilayered metasurface antenna

Initially, the MS layer is separated from the conventional chip by an air spacer, after which it is replaced by a Teflon dielectric (relative permittivity 2.1). Initially, the single-layer slotted patch antenna propagates in the TM_{01} mode. TM_{01} excitation is greatest at the power point defined by the smallest power circuit. If the MS is included in the patch in the superset configuration, the build continues in TM_{03} mode. The modal amplitude of TM_{03} is very high at the selected entry point compared to the other modes. An antenna assembly with an air gap between two FR4 dielectric layers of different thicknesses and a Teflon dielectric, shown in Figures 2.4 and

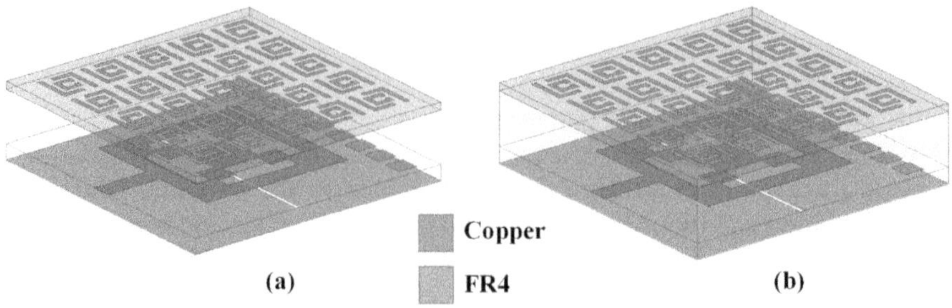

Figure 2.4 Antenna design featuring an air gap and Teflon dielectric sandwiched between two FR4 substrates [11].

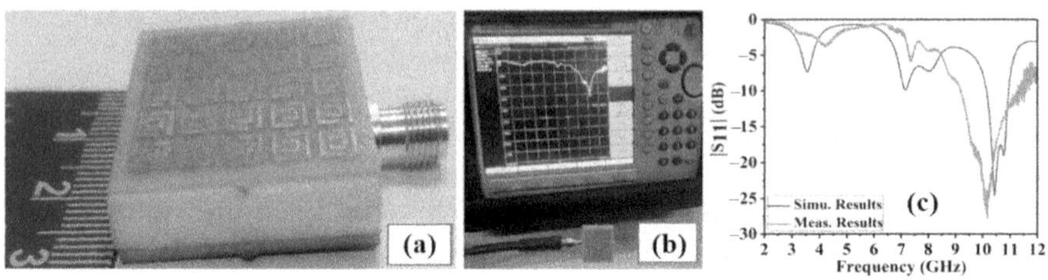

Figure 2.5 Fabricated structure: (a) 3D image of MS antenna; (b) $|S_{11}|$ measurement of VNA utilization; and (c) simulated vs. measured plot of $|S_{11}|$ (dB) in relation to frequency [11].

2.5, presents a fabricated antenna with $|S_{11}|$ measurements for using Vector Network Analyzer (VNA). MS is composed of periodic unit cells in 5×5 order. The unit cell consists of a central C-type patch, flanked by two L-type patches on the exterior. The prototype mentioned in Figure 2.6 is an left-handed material (LHM). The LHM is characterized by metal losses. To reduce the metal losses, good conductors can be used. However, it is observed that the losses are very closely related to the effective inductance (L) and effective capacitance (C) of the RLC model for LHM. To reduce the losses, increase the effective inductance-to-capacitance ratio (L/C). LHM materials are suitable for generating narrow beam radiations in far-field applications. Consequently, engineered structures can be employed to create compact, highly focused, and adaptable antennae [64–66].

The standard patch antenna has been designed with a 5×5 MS layer with air spacer. This configuration has resulted in a gain with an impedance bandwidth of 5.77% at 11.08 GHz. Subsequently, the air space between the conventional patch and the MS layer has been substituted with Teflon dielectric, as depicted in Figure 2.7. This modification has led to the achievement of an optimum gain of 8.17 dBi and a lowest frequency of 7.66% at 10.44 GHz.

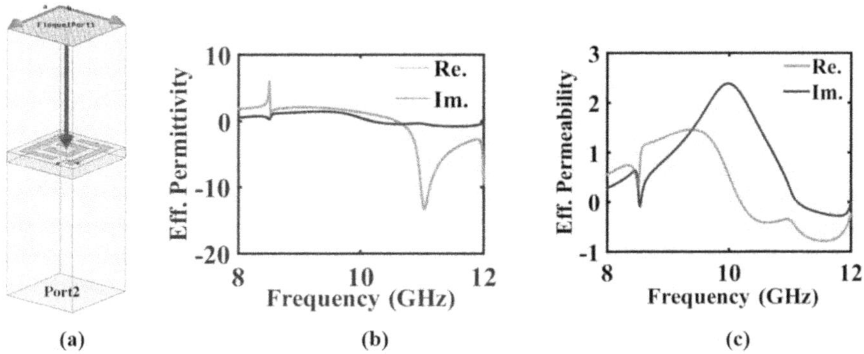

Figure 2.6 Unit cell: (a) 3D view; (b) effective permittivity; and (c) effective permeability [11].

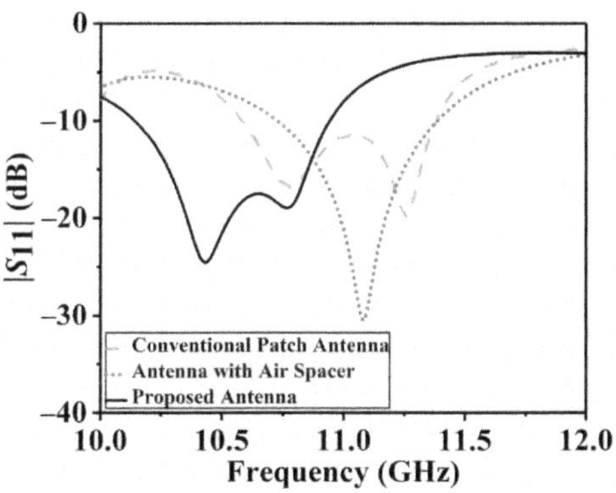

Figure 2.7 The frequency (GHz) and $|S_{11}|$ (dB) graphs for various antenna structure scenarios [11].

Figure 2.8 illustrates the prototype's far-field antenna gains to the simulated one, showing the gains with respect to frequencies. The peak directivity simulated in this setup is 8.85 dBi, with a realized gain of 8.17 dBi. On the other hand, the realized gain as measured experimentally is 7.57 dBi. As a result, the 3 dB directivity bandwidth starts at 10.07 GHz and goes up to 11.31 GHz, which is 11.8% at 10.44 GHz. It has been found that adding a MS to a straightforward conventional patch improves the antenna's gain by 7.57 dBi over the reference structure. In this chapter, the design and analysis of a multilayered dielectric MSA are discussed. The MSA employs a pair of rectangular slots within the ground plane to incorporate the flawed ground concept. This particular antenna exhibits a restricted bandwidth, resulting from feed offset and out-of-phase field leakage from certain slots

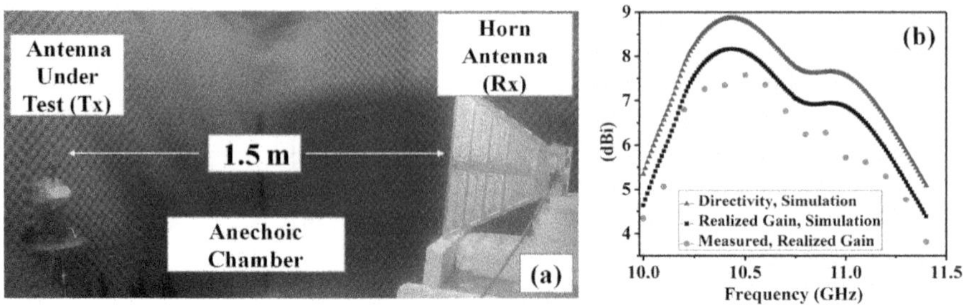

Figure 2.8 (a) Radiation pattern measurement and (b) simulated vs. measured peak gain (dB) responses versus antenna frequency [11].

within the ground plane or near the patch. Consequently, the patterns in the E- and H-planes display asymmetry and deviate toward –34°. The performance enhancement and circuit model of the multilayered antenna are further addressed in the following portion of this chapter.

2.4.2 A metasurface antenna with an air spacer: Design and circuit model analysis

The antenna design consists of a modified slotted patch and a superstrate arrangement made up of a 7 × 7 array of periodic MS [67]. The MSA is made up of a 7 × 7 order MS structure with a slotted patch that is created by square annular-shaped slots at each of the patch's four corners. Each meta-atom in the integrated MS layer is made up of a central patch encircled by four rectangular strips. The design of both the radiating patch and the MS layer was based on the FR4 dielectric layer, separated by a 2.2 mm thick air spacer. In Figure 2.9, the three-dimensional (3D) structure is illustrated, showing the typical patch and the 7 × 7 order MS superstrate layer.

An analysis is conducted on the attributes of the conventional patch antenna in a full-wave simulation and a circuit model simulation using the equivalent electrical circuit model can be seen in Figure 2.10. The series connection of parallel RLC circuits represents the equivalent circuit components of a conventional antenna [53–55]. The resulting input impedance of the analog circuit can be expressed as (9.6)

$$Z_n = \sum_{k=1}^{n} \frac{j\omega R_k L_k}{R_k\left(1-\omega^2 C_k L_k\right)+j\omega L_k} \tag{9.6}$$

Here, R_k represents the radiation resistance. L_k and C_k denote the respective inductor and capacitor values corresponding to the middle

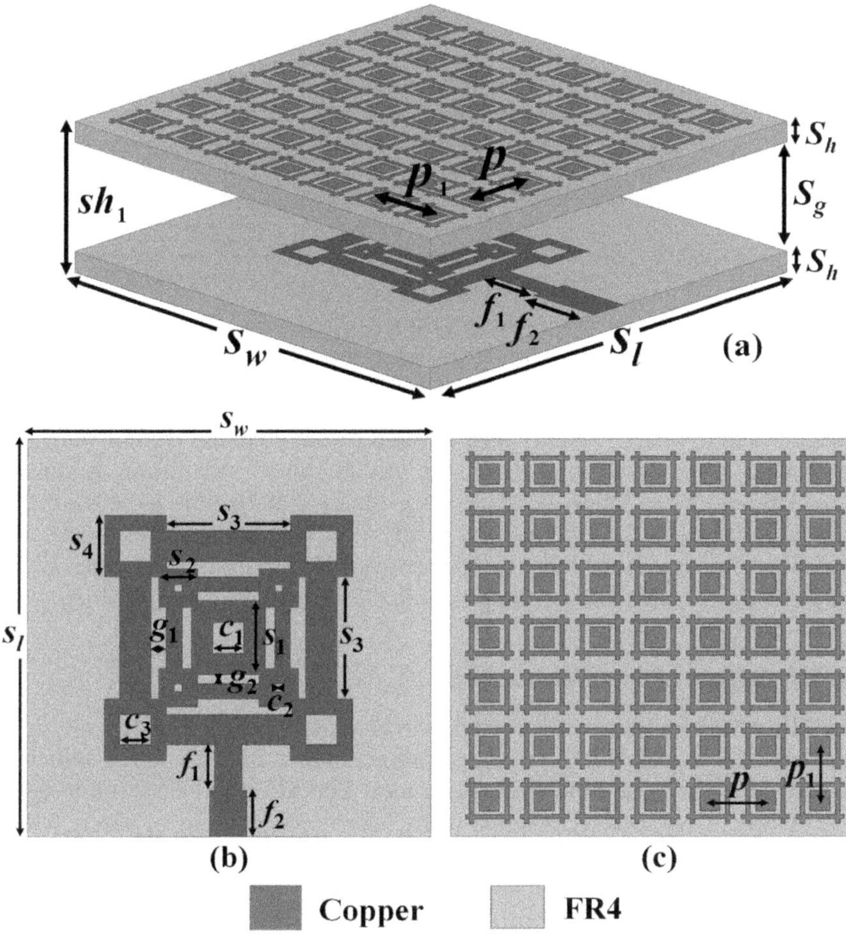

Figure 2.9 Three-dimensional antenna shape, traditional patch, the 7 × 7 order MS configuration [67].

resonator. The antenna elements are implemented with Foster's canonical expression, where R_k effectively manages both radiation and ohmic losses. RLC-parallel circuits in series as a combination of Foster's canonical form are suitable for modeling "electric antennas" that behave as an open circuit when excited by direct current, while magnetic antennas represent the parallel connection of RLC circuits in series and act as short circuits under DC excitation [53–55].

The circuit representation for a basic patch antenna that uses quarter-wave transformer feed is created by connecting parallel combinations of

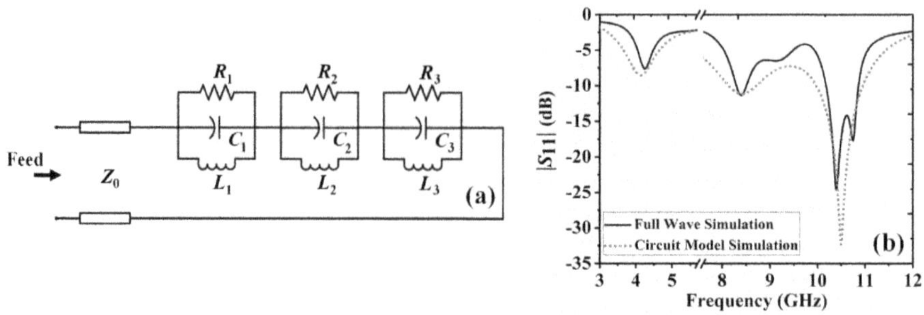

Figure 2.10 The equivalent circuit model of the patch antenna, along with its full-wave simulation and circuit model simulation characteristics [67].

RLC circuits in series [53–55]. The lumped parameters for the conventional patch have been determined from the circuit model simulation studies as $R_1 = 39.6\ \Omega$, $R_2 = 74.46\ \Omega$, $R_3 = 94.5\ \Omega$, $L_1 = 0.48$ nH, $L_2 = 0.35$ nH, $L_3 = 0.18$ nH, $C_1 = 3.38$ pF, $C_2 = 1.11$ pF, $C_3 = 1.31$ pF and $Z_0 = 50\ \Omega$. The reflection characteristics have been improved by increasing the capacitance value due to the presence of the parasitic element in the form of slots.

2.4.2.1 Design of the metasurface

The MS design is based on a periodic arrangement of 7 × 7 order unit cells. The traditional patch has been positioned below the MS layer, which is spaced apart by a 2.2 mm thick air spacer. The MS unit cell is made up of four identically shaped rectangular strips surrounding a central patch, which was built in accordance with the operational wavelength. In Figure 2.11, it can be observed that the unit cell, which is organized periodically, exhibits a reflection phase of 0° at both 10 GHz and 17.65 GHz. The fluctuation in the reflection phase is confined to a range of ±180° within the frequency range of 1–20 GHz. Below 10 GHz, the MS layer behaves as an epsilon-negative (ENG) material with $\varepsilon < 0$ and $\mu > 0$, whereas above 10 GHz, it demonstrates mu-negative (MNG) behavior with $\varepsilon > 0$ and $\mu < 0$. Hence, 10 GHz is considered as the plasma frequency as the effective permittivity is zero at 10 GHz, thereby yielding a zero-index metamaterial. The zero-phase reflection bandwidth of the MS significantly impacts the bandwidth of the array composed of unit elements. It has also been discovered that the improvement in bandwidth in the MS structures is caused by an increase in equivalent inductance [68–70].

The equivalent input impedance for MS unit elements can be represented by combining the periodic pattern impedance and a grounded dielectric layer impedance in parallel. This equivalence establishes a relationship between

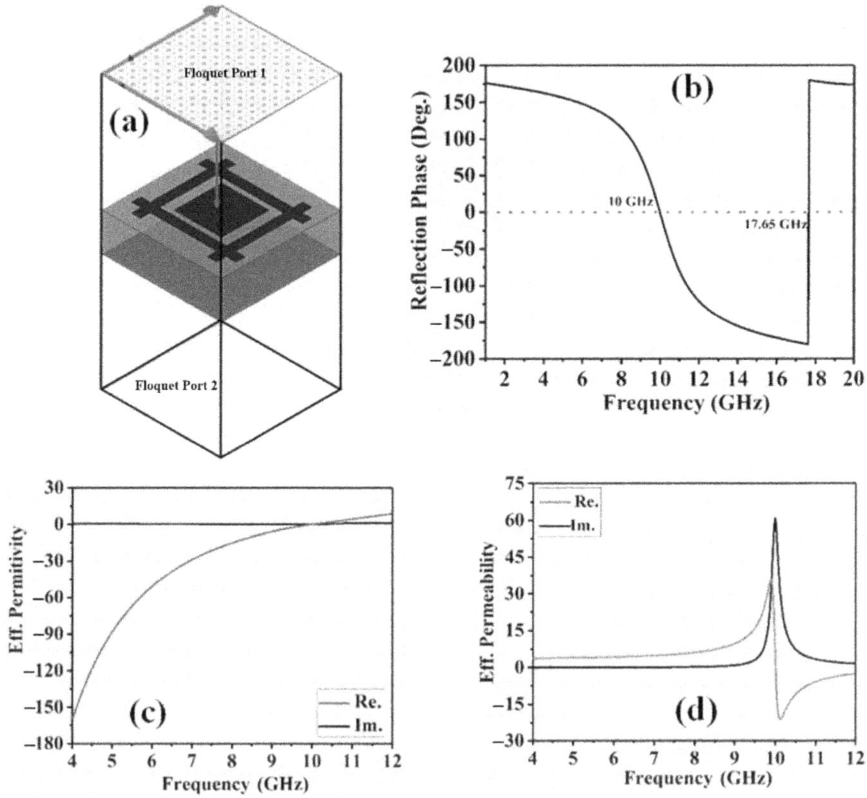

Figure 2.11 (a) The unit cell perspective; (b) considering the frequency-dependent reflection phase characteristics; (c) the effective permittivity; and (d) effective permeability of the unit cell [67].

the MS reflection bandwidth and the bandwidth of a parallel circuit with a net inductance L_m and capacitance C_m, as illustrated in equation (9.7):

$$BW=\frac{\omega_0^2 L_m}{Z_0}=\frac{1}{C_m Z_0} \tag{9.7}$$

where $\omega_0 L_m$ represents the impedance, ω_0 which has a dimension of rad/sec represents the bandwidth, and Z_0 represents the impedance of free space. By introducing a wide slot gap between the centrally positioned patch and the rectangular strips on each arm of the MS unit cell, it is possible to enhance the zero-phase bandwidth of the MS structure by reducing C_m. The capacitive impact of the etched slot gap between the centrally positioned patch and the rectangular strips of the unit cell decreases the overall equivalent capacitance of the MS structure when these capacitances are connected in series, thereby expanding the bandwidth for zero-phase reflection.

In this MS antenna design, an arrangement of periodic unit cells in a 7 × 7 array has been implemented as a superstrate configuration. The superstrate is positioned above the modified patch antenna, with a separation distance of Sh_1 from the ground layer. The resonant conditions of this MS antenna, which relies on the superstrate, can be described within the range of (9.8–9.9) [70–72]

$$\varphi_s + \varphi_g - \frac{4\pi h}{\lambda} = 2N\pi, \qquad N = 0,\ \pm 1,\ \pm 2, \ldots \tag{9.8}$$

$$f = \frac{c}{4\pi h}\left(\varphi_s + \varphi_g - 2N\pi\right) \tag{9.9}$$

where h is the distance that separates the ground from the superstrate layer. The reflection phases of the superstrate layer are designated as φ_s, while the reflection phases of the metal ground plane are denoted as φ_g. The operating wavelength in the substrate is represented by λ, differing from the free space wavelength λ_0 in the conventional air-filled antenna. As λ is shorter than λ_0, as observed from (9.8), the height h of the MS antenna is smaller compared to the conventional air-filled one.

As shown in Figure 2.12, the equivalent circuit model of the MS superstrate-loaded antenna has since been developed. The air gap layer between the patch and the MS has been addressed with the boundaries R_a and C_a while R_m, C_m, and L_m are the same lumped components utilized for unit cells. Furthermore, the radiation resistance of the MS resonator is discussed by R_m. L_m and C_m represent the specific inductance and capacitance values of the MS space resonator. As stated earlier, R_m in the Foster canonical form representation of the MS accounts for both radiation and ohmic losses. C_{cm1} and C_{cm2} denote the coupling capacitances between adjacent unit cells in the vertical and horizontal directions of the MS layer. Based on Keysight Advanced Design System (ADS) simulation, the following parameters were calculated for the lumped circuit: R_a = 377, C_a = 220.65 pF, R_m = 94.5, C_m = 1.186 pF, L_m = 0.273 nH, C_{cm1} = 0.35 pF, C_{cm2} = 0.35 pF and Z_0 = 50. The additional circuit component values for the MS as a superstrate have been derived by fixing the same circuit component values of the radiating patch.

Figure 2.13 shows the reflection coefficients characteristics of an MS antenna, along with that of an outlying patch antenna. With a fractional bandwidth of 2.4% and 6.5%, the radiating patch operates in two frequency

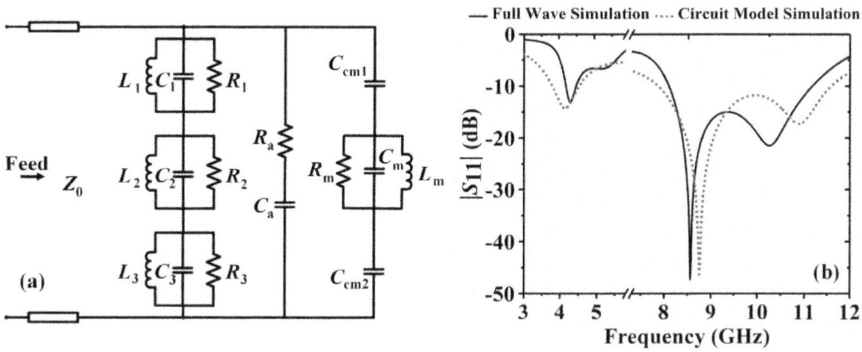

Figure 2.12 (a) The MS antenna equivalent electrical circuit model; (b) comparison of the characteristics of the ADS simulation and the Ansys HFSS simulation for the plot of $|S_{11}|$ (dB) with respect to frequency [67].

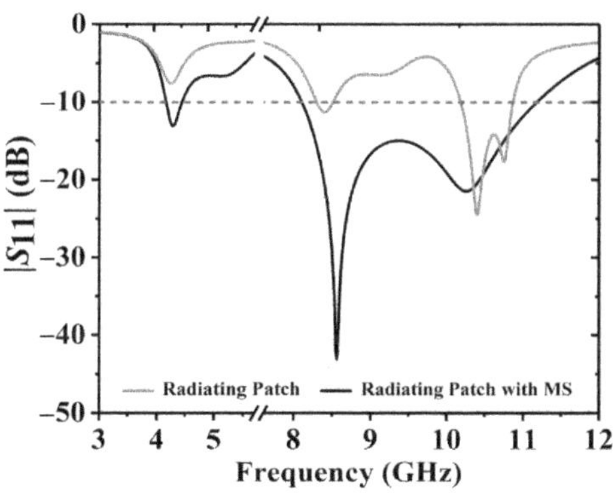

Figure 2.13 $|S_{11}|$ (dB) vs. frequency (GHz) of the antenna [67].

bands from 8.3 to 8.51 GHz and from 10.19 to 10.87 GHz. This patch antenna exhibits the narrowband behavior of impedance bandwidth with low return loss. The performance of this radiating patch has been significantly improved by incorporating a 7 × 7 order MS layer with air spacer in the superstrate configuration. The antenna is able to operate within the frequency ranges of 4.18–GHz and 8.11–11.14 GHz by incorporating the MS superstrate layer, leading to a dual-band response. The fractional bandwidths at 4.26 GHz and 8.62 GHz are measured at 6.1% and 35.15% respectively. A maximum return loss of 42dB was observed at 8.62 GHz.

Figure 2.14 shows the simulated and measured radiation patterns of the antenna along the E-plane ($\varphi = 0^0$) and H-plane ($\varphi = 90^0$) at different operating frequencies, respectively. The MS antenna radiates in the boresight

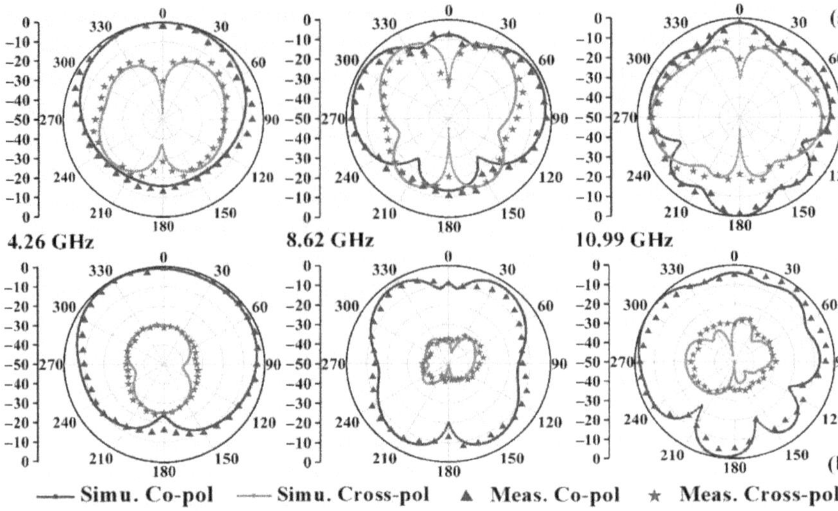

Figure 2.14 Co-polarized and cross-polarized radiation characteristics of the MS antenna in the (a) E-plane and (b) H-plane [67].

direction in both E- and H-planes across all operational frequencies. However, at 8.62 GHz, the antenna pattern deviates toward ±30°.

2.4.3 A metasurface antenna with solid spacer: Design and circuit model analysis

There are several monopole antennas that have been extensively used with the MS surface to reduce the antenna profile and improve device performance [73–77]. The MS-based low-profile dual-polarized, wideband pattern diversity antenna is a suitable choice for in-band full-duplex applications. In wireless applications, the co-planar waveguide (CPW)-fed slot antenna offers significant advantages, especially with its multiband or wideband capabilities. Nevertheless, it is important to mention that this antenna has low gain [76–78]. To overcome the gain limitations of this antenna type, the MS with a radiating element proves to be beneficial in addressing gain issues at high frequencies and also proven effective in reducing radar cross-section [78–83].

In this setup, a slotted patch antenna is connected via CPW and paired with a 7 × 7 order MS surface, with a layer of Teflon in between. Teflon offers beneficial electrical properties, enhances the antenna's mechanical strength, and showcases resilience to changes in temperature. The radiating patch is placed over the initial dielectric, while the MS surface is positioned on top of the subsequent dielectric layer. Additionally, to validate the outcomes of

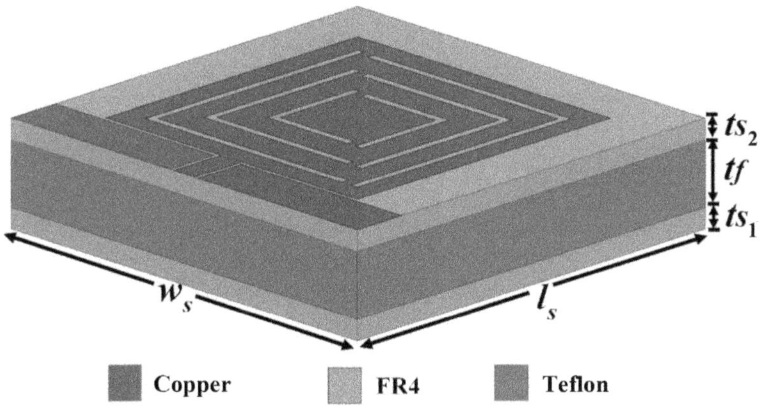

Figure 2.15 The three-dimensional perspective of the metasurface configuration [79].

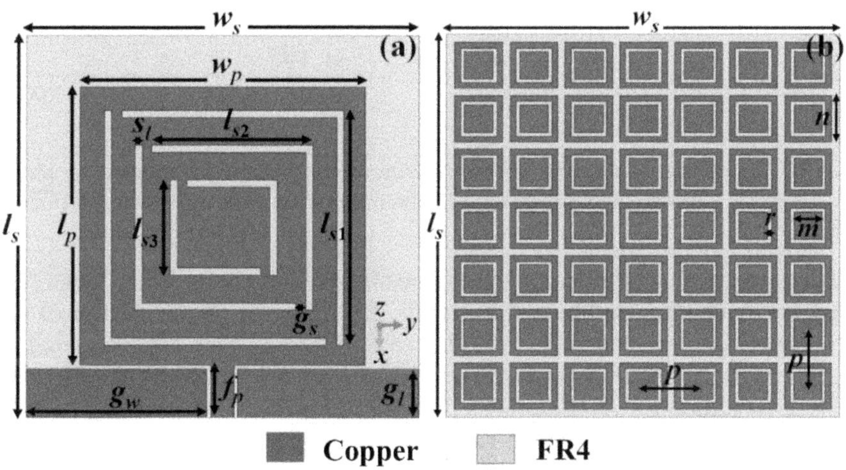

Figure 2.16 The MS antenna is equipped with both a radiating patch and a 7 × 7 order MS surface [79].

the EM simulation, the equivalent circuit model of the suggested prototype has been taken into account.

The antenna is made up of a slotted patch that is fed by CPW and a 7 × 7 order MS. The patch and MS reflector are built on two distinct FR4 dielectric substrates, with a Teflon layer measuring 5 mm in thickness placed in between. Figure 2.15 illustrates a three-dimensional schematic depiction of the MS structure, whereas Figure 2.16 showcases both the radiating patch and the 7 × 7 ordered MS surface of the antenna.

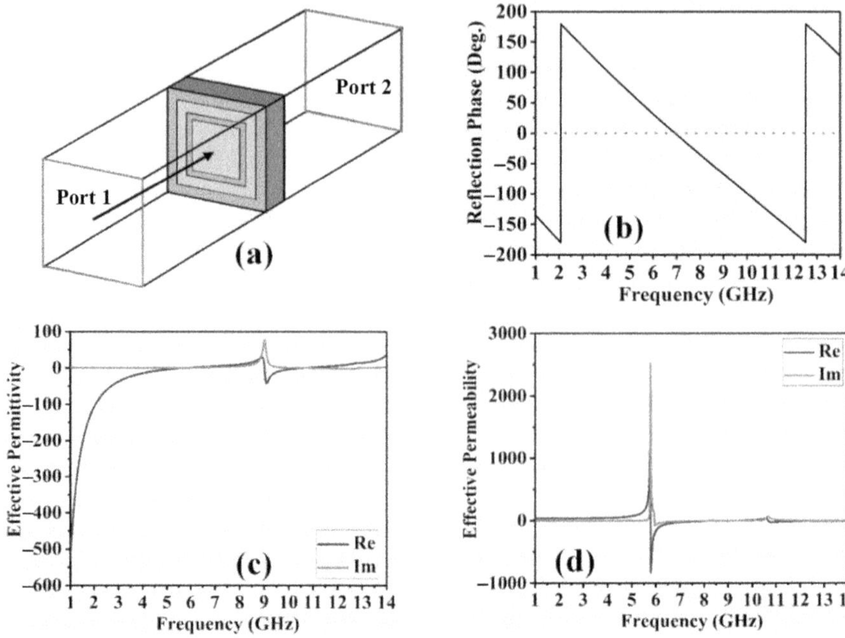

Figure 2.17 (a) Model of the unit cell for analysis; (b) characteristics of reflection phase; (c) effective permittivity; and (d) effective permeability of the unit cell [79].

The arrangement of MS unit cell can be observed in Figure 2.17, displaying a central patch encircled by a square ring. By adjusting the properties of the MS unit cell, an equivalent LC parallel resonance has been established. The designs of the unit cell are tailored to various real-world uses. The reflection phase characteristics of the unit cell can be observed in Figure 2.17, indicating that the reflection phase varies between ±180° across the frequency range of 2–12.5 GHz. The zero-crossing reflection bandwidth of the MS unit cell plays a crucial role in determining the bandwidth of the periodic structures. Furthermore, a detailed analysis of the reflection and transmission properties of the unit cell has been conducted to thoroughly investigate the effective permittivity and permeability of the MS layer [64–66]. The frequency range of 1–3.8 GHz and 8.8–11 GHz is where the MS layer functions as a left-handed material. During the design of the MS's unit cells, the homogeneity of the MS has been considered.

Furthermore, the radiating element is illustrated as a combination of multiple parallel RLC circuits interconnected in a sequential manner within the equivalent electrical circuit diagram of the MSA, as depicted in Figure 2.18.

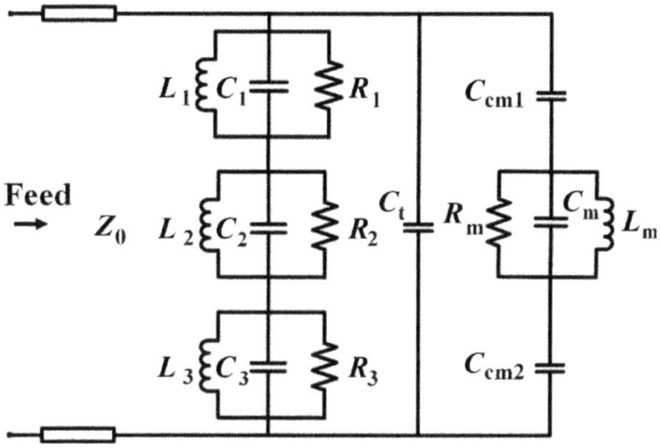

Figure 2.18 A comparable circuit model can be employed as a representation of the configuration [79].

A circuit model for the initial set of slots have been constructed using a parallel configuration of R_1, L_1, and C_1. The central slotted region and the outer pair of slots of the radiating patch were represented by the parallel combinations of R_2, L_2, C_2, R_3, L_3, and C_3. The presence of a Teflon layer between the patch and the MS on the lower substrate was symbolized by the parameter C_t, which was connected in parallel with the antenna circuit. The equivalent lumped elements for the unit cells of the MS, namely—R_m, C_m, and L_m, were used. R_m specifically represented the radiation resistance of the MS resonator. The MS slot resonator was characterized by its equivalent inductance (L_m) and capacitance (C_m) values. The R_m component accounted for both radiation and ohmic losses in the resonator. Additionally, the vertical and horizontal coupling capacitances between adjacent unit cells in the MS layer were denoted as C_{cm1} and C_{cm2}, respectively.

The inductance L_m and the capacitor C_m can be estimated using the equations (9.10, 9.11) [55, 78]

$$L_m = \mu h \tag{9.10}$$

$$C_m = \frac{n\varepsilon_0\left(1+\varepsilon_r\right)}{\pi}\cosh^{-1}\left(\frac{2n+g}{g}\right) \tag{9.11}$$

The substrate's height is symbolized as h, the unit cell patch's width is represented by n, and g stands for the gap between two adjacent unit cells. The mathematical expression for the bandwidth of the in-phase reflection of the MS is given by (9.12)

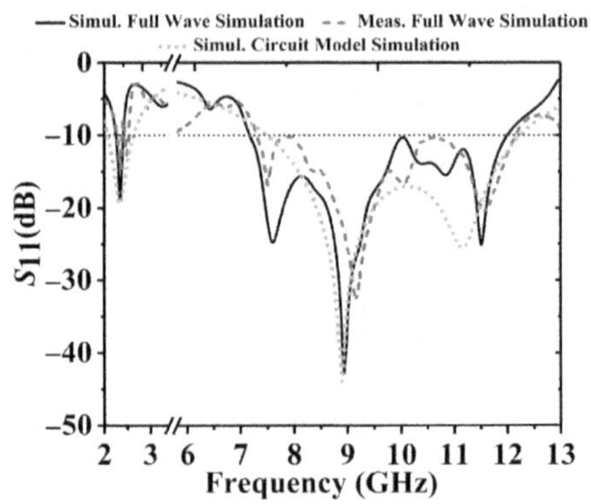

Figure 2.19 Comparison of the characteristics between full wave simulation and circuit model simulation reveals the variation in the plot of S_{11} (dB) with respect to frequency [79].

$$BW = \frac{1}{\eta}\sqrt{\frac{L_m}{C_m}} \tag{9.12}$$

The input impedance of the interconnected circuit, which consists of the radiating patch, Teflon layer, and MS layer, may be expressed using equations (9.6), (9.13), and (9.14)

$$Z_t = \frac{1}{j\omega C_t} \tag{9.13}$$

$$Z_m = \frac{1}{j\omega C_{cm1}} + \frac{j\omega R_m L_m}{R_m\left(1-\omega^2 C_m L_m\right)+j\omega L_m} + \frac{1}{j\omega C_{cm2}} \tag{9.14}$$

The lumped circuit parameters shown in Figure 2.18 have been derived through simulation analysis using Keysight ADS [25–28] and these are: R_1 = 43.7064 Ω, R_2 = 64.7883 Ω, R_3 = 77.1983 Ω, L_1 = 1.2035 nH, L_2 = 0.5123 nH, L_3 = 0.1942 nH, C_1 = 4.2023 pF, C_2 = 0.7428 pF, C_3 = 0.9015 pF, C_t = 0.0371 pF, R_m = 133.747 Ω, C_m = 3.3426 pF, L_m = 1.2542 nH, C_{cm1} = 0.001 pF, C_{cm2} = 259.922 pF and Z_0 = 50 Ω. Figure 2.19 illustrates the comparison of the reflection coefficient responses between the circuit model simulation and the numerical simulation, demonstrating a satisfactory similarity between the two. The prototype is designed to operate in two distinct frequency bands: 2.25–2.41 GHz and 7.27–11.98 GHz, with the

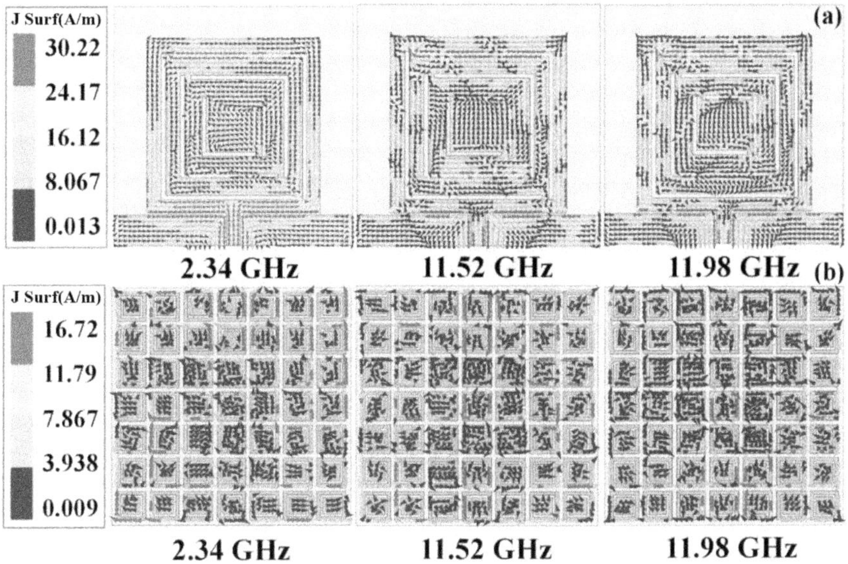

Figure 2.20 Surface current distributions for the radiating patch and 7 × 7 order MS change depending on operating frequency [79].

latter encompassing the *X*-band region. A dual-band antenna has the capability to transmit and receive radiofrequency (RF) signals at two different frequencies. These antennas can utilize either frequency individually or simultaneously, depending on their configuration. Dual-band antennas offer exceptional performance and versatility. Some devices can connect to one frequency, while others can connect to a different frequency to achieve improved speed and range. The antenna used operates at frequencies of 2.34 GHz and 8.94 GHz, with fractional bandwidths of 6.8% and 53.8% respectively. Additionally, at 8.94 GHz, a maximum return loss of 42 dB has been achieved. Typically, a patch antenna resonates at a specific frequency. However, in this case, the dual-band functionality is achieved through the combined impact of slot impedance in the patch and the planar MS. The planar MS structure acts as a high impedance surface that can modulate the EM fields generated by the patch resonator.

The influence of slots and the MS resonances on antenna performance can be evaluated by studying the surface current distributions of the radiating patch and MS layer. In Figure 2.20, the surface current distributions of the traditional patch and the MS are shown at different operating frequencies. At 2.34 GHz, there is a notable coupling current in both the radiating patch and the MS. However, at 11.52 GHz and 11.98 GHz, the coupling currents

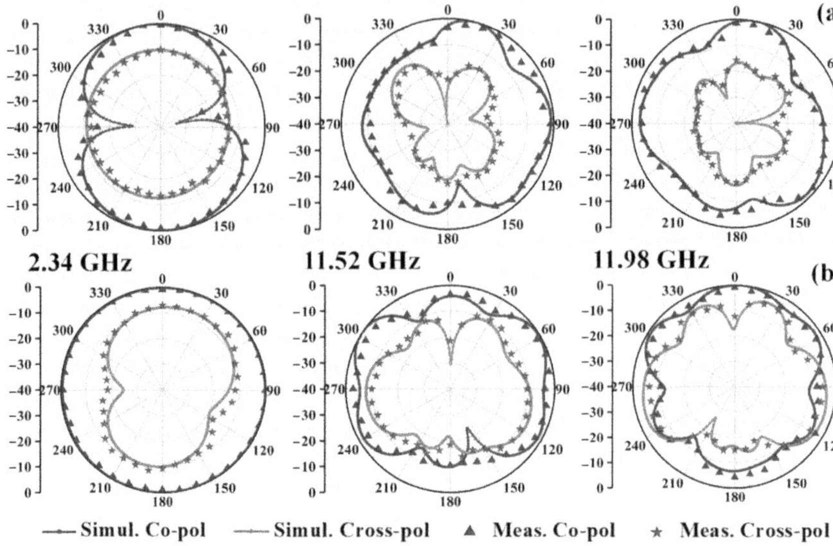

Figure 2.21 The radiation patterns in the E-plane and H-plane at frequencies of 4.13 GHz, 8.80 GHz, and 12.01 GHz [79].

in the patch slots and MS unit cells decrease. Initially, the surface area of the radiating patch displays strong surface current distributions. With the introduction of the MS layer, the coupling current on the antenna slots and the region between the patch surface contact and the CPW feed is significantly reduced, leading to the elimination of remaining edge currents. Conversely, at higher frequencies, the presence of fewer coupling currents improves the antenna's efficiency, allowing it to transmit maximum power.

The radiation patterns of the MS antenna are depicted in Figure 2.21 along the E- and H-planes. This antenna demonstrates directional radiation characteristics in both planes across all operating frequencies. Moreover, the antenna's cross-polarized levels consistently register lower than the co-polarized levels at the operational frequencies. The radiation patterns exhibit variations across the operational bands, possibly due to the leakage of out-of-phase fields from a few slots at the patch, ground slots, or the unit cells of the MS. The presence of coupling currents in the device may be a consequence of the close spacing between the unit cells and the patch. Nonetheless, this distortion at high frequencies remains within acceptable limits for wireless applications in multipath environments, as the antenna's cross-polarized levels continue to be lower than the co-polarized levels.

2.5 CONCLUSION

Metasurface antennas offer unique capabilities for various applications, including communication, sensing, and radar. Through careful design, optimization, and integration, these antennas can address the challenges of modern wireless systems and pave the way for advanced functionalities in future wireless technologies. MSA are advantageous due to their small size, directed main lobe, and flexibility in interfacing with multiple systems. Further research is needed to optimize the integration of MSA with other system components, such as RF front-ends and signal processing units.

REFERENCES

[1] J. Zhang, S. Yan, and G. A. E. Vandenbosch, "Radial CRLH-TL-Based Dual-Band Antenna with Frequency Agility," *IEEE Transactions on Antennas and Propagation*, vol. 68, no. 7, pp. 5664–5669, July 2020.

[2] H. Li, G. Wang, X. Gao, J. Liang, and H. Hou, "A Novel Metasurface for Dual-Mode and Dual-Band Flat High-Gain Antenna Application," *IEEE Transactions on Antennas and Propagation*, vol. 66, pp. 3706–3711, July 2018.

[3] J. Canet-Ferrer, Metamaterials and Metasurfaces, IntechOpen, UK, chapter 4, pp. 63–85, 2019. ISBN: 978-1-83881-735-0.

[4] Y. Wang and A. Zhang, "Dual Circularly Polarized Fabry–Perot Resonator Antenna Employing a Polarization Conversion Metasurface," *IEEE Access*, vol. 9, pp. 44881–44887, 2021.

[5] C. Holloway, A. Dienstfrey, E. F. Kuester, J. F. O'Hara, A. K. Azad, and A. J. Taylor, "A Discussion on the Interpretation and Characterization of Metafilms/Metasurfaces: The Two-Dimensional Equivalent of Metamaterials," *Metamaterials*, vol. 3, no. 2, pp.100–112, October 2009.

[6] Y. Zheng, J. Gao, Y. Zhou, X. Cao, H. Yang, S. Li, and T. Li, "Wideband Gain Enhancement and RCS Reduction of Fabry–Perot Resonator Antenna with Chessboard Arranged Metamaterial Superstrate," *IEEE Transactions on Antennas and Propagation*, vol. 66, no. 2, pp. 590–599, February 2018.

[7] Y. Dong and T. Itoh, "Metamaterial-Based Antennas," *Proceedings of the IEEE*, vol. 100, no. 7, pp. 2271–2285, July 2012.

[8] C. L. Holloway, E. F. Kuester, J. A. Gordon, J. O'Hara, J. Booth, and D. R. Smith, "An Overview of the Theory and Applications of Metasurfaces: The Two-Dimensional Equivalents of Metamaterials," *Antennas and Propagation Magazine, IEEE*, vol. 54, pp. 10–35, 2012.

[9] R. Nakamura and T. Fukusako, "Broadband Design of Circularly Polarized Microstrip Patch Antenna Using Artificial Ground Structure with Rectangular Unit Cells," *IEEE Transactions on Antennas and Propagation*, vol. 59, 2103, 2011.

[10] D. Samantaray, S. Bhattacharyya, and K. V. Srinivas, "A Modified Fractal-shaped Slotted Patch Antenna with Defected Ground Using Metasurface for Dual Band Applications," *International Journal of RF and Microwave Computer-Aided Engineering*, vol. 29, no. 12, pp. 1–9, December 2019.

[11] D. Samantaray and S. Bhattacharyya, "A Gain-Enhanced Slotted Patch Antenna Using Metasurface as Superstrate Configuration," *IEEE Transactions on Antennas and Propagation*, vol. 68, no. 9, pp. 6548–6556, September 2020.
[12] Q. Liu, H. Liu, W. He, and S. He, "A Low-Profile Dual-Band Dual-Polarized Antenna with an AMC Reflector for 5G Communications," *IEEE Access*, vol. 8, pp. 24072–24080, January 2020.
[13] M. Chandra, Nilotpal, D. Samantaray, M. Thottappan, and S. Bhattacharyya, "A Broadband Transmissive Type Metasurface Cross-Polarization Converter for EMC Application," *IEEE Transactions on Electromagnetic Compatibility*, vol. 65, no. 1, pp. 186–194, February 2023.
[14] F. Costa, O. Luukkonen, C. R. Simovski, A. Monorchio, S. A. Tretyakov, and P. M. de Maagt, "TE Surface Wave Resonances on High-Impedance Surface-Based Antennas: Analysis and Modeling," IEEE Transactions on Antennas and Propagation, vol. 59, no. 10, pp. 3588–3596, October 2011
[15] I. Park, "Application of Metasurfaces in the Design of Performance-Enhanced Low-Profile Antennas," *EPJ Applied Metamaterials*, vol. 5, 11, 2018.
[16] S.X. Ta and I. Park, "Artificial Magnetic Conductor Based Circularly Polarized Crossed-Dipole Antennas _ 1. AMC Structure with Grounding Pins," *Radio Science*, vol. 52, 630, 2017.
[17] S.X. Ta, I. Park, "Artificial Magnetic Conductor Based Circularly Polarized Crossed-Dipole Antennas _ 2. AMC Structure Without Grounding Pins," *Radio Science*, vol. 52, 642, 2017.
[18] D. Samantaray and S. Bhattacharyya, "A Metasurface Based Gain Enhanced Dual Band Patch Antenna Using SRRs with Defected Ground Structure," *Radio Science*, vol. 56, no. 2, pp. 1–13, article no. 2020RS007192, February 2021.
[19] C. A. Balanis, *Antenna Theory Analysis and Design*, Third Edition, John Willy and Sons, Inc. Wiley Reprint, India, 2010.
[20] D. M. Pozar and D. H. Schaubert, *Microstrip Antennas*, IEEE Press, New York, 1995.
[21] C. Saha, J.Y. Siddiqui, and Y. Antar, *Multifunctional Ultra Wideband Antennas: Trends, Techniques and Applications*, Taylor and Francis, Boca Raton, FL, 2019.
[22] D. Samantaray and S. Bhattacharyya, "An AMC Based Metasurface Patch Antenna for C-Band and X-Band Applications," *IEEE Indian Conference on Antennas and Propagation (InCAP 2021)*, pp. 548–551, Jaipur, India, December 13–16, 2021.
[23] R. O. Ouedraogo, E. J. Rothwell, A. R. Diaz, K. Fuchi, and A. Temme, "Miniaturization of Patch Antennas Using a Metamaterial-Inspired Technique," IEEE Transactions on Antennas and Propagation, vol. 60, no. 5, pp. 2175–2182, May 2012.
[24] N. Zhu and R. W. Ziolkowski, "Broad-Bandwidth, Electrically Small Antenna Augmented with an Internal Non-Foster Element," *IEEE Antennas and Wireless Propagation Letters*, vol. 11, pp. 1116–1120, 2012.

[25] L. M. Si, W. Zhu, and H. J. Sun, "A Compact, Planar, and CPW-Fed Metamaterial-Inspired Dual-Band Antenna," IEEE Antennas and Wireless Propagation Letters, vol. 12, pp. 305–308, 2013.
[26] D. Samantaray and S. Bhattacharyya, "A Gain Enhanced Metasurface based Monopole Antenna for WLAN Application," accepted for presentation in URSI General Assembly 2021, Rome, Italy, August 28–September 4, 2021.
[27] H. Huang, Y. Liu, S. Zhang, and S. Gong, "Multiband Metamaterial-Loaded Monopole Antenna for WLAN/WiMAX Applications," IEEE Antennas and Wireless Propagation Letters, vol. 14, pp. 662–665, 2015.
[28] J. C. Myers, P. Chahal, E. Rothwell, and L. Kempel, "A Multilayered Metamaterial-Inspired Miniaturized Dynamically Tunable Antenna," *IEEE Transactions on Antennas and Propagation*, vol. 63, no. 4, pp. 1546–1553, April 2015.
[29] A. H. Naqvi and S. Lim, "A Beam-Steering Antenna with a Fluidically Programmable Metasurface," *IEEE Transactions on Antennas and Propagation*, vol. 67, no. 6, pp. 3704–3711, June 2019.
[30] G. Von Trentini, "Partially Reflecting Sheet Arrays," *IRE Transactions on Antennas and Propagation*, vol. 4, no. 4, pp. 666–671, October 1956.
[31] D. H. Kwon, "Lossless Scalar Metasurfaces for Anomalous Reflection Based on Efficient Surface Field Optimization," *IEEE Antennas and Wireless Propagation Letters*, vol. 17, no. 7, pp. 1149–1152, July 2018.
[32] O. Quevedo-Teruel, M. Ebrahimpouri, and F. Ghasemifard, "Lens Antennas for 5G Communications Systems," *IEEE Communications Magazine*, vol. 56, no. 7, pp. 36–41, July 2018.
[33] M. D. Gregory, J. A. Bossard, Z. C. P. O. Morgan, C. S. Cicero, J. A. Easum, J. D. Binion, D. Z. Zhu, C. P. Scarborough, P. L. Werner, D. H. Werner, S. Griffiths, M. Ketner, and J. Pompeii, "A Low Cost and Highly Efficient Metamaterial Reflector Antenna," IEEE Transactions on Antennas and Propagation, vol. 66, no. 3, pp. 1545–1548, March 2018.
[34] M. Li, Y. Zhang, and M. Tang, "Design of a Compact, Wideband, Bidirectional Antenna Using Index-Gradient Patches," *IEEE Antennas and Wireless Propagation Letters*, vol. 17, no. 7, pp. 1218–1222, July 2018.
[35] D. Samantaray, S. Bhattacharyya, and K. V. Srinivas, "A Gain Enhanced Multiband Antenna using SRRs with Defected Ground Structure," *2019 URSI Asia Pacific Radio Science Conference (AP-RASC 2019)*, New Delhi, India, March 9–15, 2019.
[36] S. Zarbakhsh, M. Akbari, F. Samadi, and A. Sebak, "Broadband and High-Gain Circularly-Polarized Antenna with Low RCS," IEEE Transactions on Antennas and Propagation, vol. 67, no. 1, pp. 16–23, January 2019.
[37] W. Yang, S. Chen, Q. Xue, W. Che, G. Shen, and W. Feng, "Novel Filtering Method Based on Metasurface Antenna and Its Application for Wideband High-Gain Filtering Antenna with Low Profile," IEEE Transactions on Antennas and Propagation, vol. 67, no. 3, pp. 1535–1544, March 2019.
[38] K. Liu, Y. Ge, and C. Lin, "A Compact Wideband High-Gain Metasurface-Lens-Corrected Conical Horn Antenna," IEEE Antennas and Wireless Propagation Letters, vol. 18, no. 3, pp. 457–461, March 2019.

[39] T. Yue, Z. H. Jiang, and D. H. Werner, "A Compact Metasurface-Enabled Dual-Band Dual-Circularly Polarized Antenna Loaded with Complementary Split Ring Resonators," IEEE Transactions on Antennas and Propagation, vol. 67, no. 2, pp. 794–803, Feb. 2019.
[40] Y. Juan, W. Yang, and W. Che, "Miniaturized Low-Profile Circularly Polarized Metasurface Antenna Using Capacitive Loading," IEEE Transactions on Antennas and Propagation, vol. 67, no. 5, pp. 3527–3532, May 2019.
[41] Y. M. Pan, P. F. Hu, X. Y. Zhang, and S. Y. Zheng, "A Low-Profile High-Gain and Wideband Filtering Antenna with Metasurface," IEEE Transactions on Antennas and Propagation, vol. 64, no. 5, pp. 2010–2016, May 2016.
[42] M. J. Al-Hasan, T. A. Denidni, and A. R. Sebak, "Millimeter-Wave EBG-Based Aperture-Coupled Dielectric Resonator Antenna," IEEE Transactions on Antennas and Propagation, vol. 61, no. 8, pp. 4354–4357, August 2013.
[43] Z. Guo, H. Tian, X. Wang, Q. Luo, and Y. Ji, "Bandwidth Enhancement of Monopole UWB Antenna with New Slots and EBG Structures," IEEE Antennas and Wireless Propagation Letters, vol. 12, pp. 1550–1553, 2013.
[44] T. H. Vu, S. Collardey, A. C. Tarot, and K. Mahdjoubi, "Input Impedance of Fabry-Perot, EBG, and Leaky-Wave Antennas Excited by a Line Source," IEEE Antennas and Wireless Propagation Letters, vol. 8, pp. 248–252, 2009.
[45] J. de D. Ntawangaheza, L. Sun, C. Yang, Y. Pang, and G. Rushingabigwi, "Thin-Profile Wideband and High-Gain Microstrip Patch Antenna on a Modified AMC," IEEE Antennas and Wireless Propagation Letters, vol. 18, no. 12, pp. 2518–2522, December 2019.
[46] S. Shi, P. Yang, W. Feng, L. Zhou, Q. Lu, W. Chen, and W. Che, "Wideband Planar Phased Array Antenna Based on Artificial Magnetic Conductor Surface," IEEE Transactions on Circuits and Systems II: Express Briefs, vol. 67, no. 10, pp. 1909–1913, October 2020.
[47] H. Wu, J. Geng, K. Wang, H. Zhou, X. Zhao, C. He, X. Liang, W. Zhu, R. Jin, "A Low-Profile Wideband Dual-Polarized Antenna Based on an Improved HIS and Its Broad-Angle Beam-Scanning Array," IEEE Antennas and Wireless Propagation Letters, vol. 19, no. 3, pp. 383–387, March 2020.
[48] N. M. Mohamed-Hicho, E. Antonino-Daviu, M. Cabedo-Fabrés, and M. Ferrando-Bataller, "A Novel Low-Profile High-Gain UHF Antenna Using High-Impedance Surfaces," IEEE Antennas and Wireless Propagation Letters, vol. 14, pp. 1014–1017, 2015.
[49] J. de D. Ntawangaheza, L. Sun, Y. Li, D. Biao, Z. Xie, and G. Rushingabigwi, "A Single-Layer Planar Low-Profile Wideband Microstrip Line-Fed Metasurface Antenna," IEEE Antennas and Wireless Propagation Letters, vol. 20, no. 9, pp. 1641–1645, Sept. 2021.
[50] F. Costa, A. Monorchio, and G. Manara, "An Overview of Equivalent Circuit Modeling Techniques of Frequency Selective Surfaces and Metasurfaces," *Applied Computational Electromagnetics Society Journal*, vol. 29, no. 12, 2014.

[51] F. Costa, A. Monorchio, and G. Manara, "Efficient Analysis of Frequency-Selective Surfaces by a Simple Equivalent Circuit Model," *IEEE Antennas and Propagation Magazine*, vol. 54, no. 4, pp. 35–48, August 2012.
[52] M. Borgese and F. Costa, "A Simple Equivalent Circuit Approach for Anisotropic Frequency-Selective Surfaces and Metasurfaces," *IEEE Transactions on Antennas and Propagation*, vol. 68, no. 10, pp. 7088–7098, 2020.
[53] S. B. T. Wang, A. M. Niknejad, and R. W. Brodersen, "Circuit Modeling Methodology for UWB Omnidirectional Small Antennas," *IEEE Journal on Selected Areas in Communications*, vol. 24, no. 4, pp. 871–877, April 2006.
[54] K. H. Sayidmarie and L. S. Yahya, "Modeling of Dual-Band Crescent-Shape Monopole Antenna for WLAN Applications," *International Journal of Electromagnetics and Applications*, vol. 4, No. 2, pp. 31–39, 2014.
[55] F. Yu and C. Wang, "Design of a CPW-Fed Dual Band-Notched planar Wideband Antenna for UWB Applications," In M. A. Matin, *Ultra Wideband Communications*, IntechOpen, UK, pp. 239–254, August 2011.
[56] H. Zhai, K. Zhang, S. Yang, and D. Feng, "A Low-Profile Dual-Band Dual-Polarized Antenna with an AMC Surface for WLAN Applications," IEEE Antennas and Wireless Propagation Letters, vol. 16, pp. 2692–2695, 2017.
[57] J. Wu, S. Yang, Y. Chen, S. Qu, and Z. Nie, "A Low Profile Dual-Polarized Wideband Omnidirectional Antenna Based on AMC Reflector," IEEE Transactions on Antennas and Propagation, vol. 65, no. 1, pp. 368–374, January 2017.
[58] Y. F. Cao and X. Y. Zhang, "A Wideband Beam-Steerable Slot Antenna Using Artificial Magnetic Conductors with Simple Structure," *IEEE Transactions on Antennas and Propagation*, vol. 66, no. 4, pp. 1685–1694, April 2018.
[59] J. Wang, H. Wong, Z. Ji, and Y. Wu, "Broadband CPW-Fed Aperture Coupled Metasurface Antenna," IEEE Antennas and Wireless Propagation Letters, vol. 18, no. 3, pp. 517–520, March 2019.
[60] K. Konstantinidis, A. P. Feresidis, and P. S. Hall, "Broadband Sub-Wavelength Profile High-Gain Antennas Based on Multi-Layer Metasurfaces," IEEE Transactions on Antennas and Propagation, vol. 63, no. 1, pp. 423–427, January 2015.
[61] J. H. Kim, C. Ahn, and J. Bang, "Antenna Gain Enhancement Using a Holey Superstrate," IEEE Transactions on Antennas and Propagation, vol. 64, no. 3, pp. 1164–1167, March 2016.
[62] Y. Liu and X. Zhao, "Perfect Absorber Metamaterial for Designing Low-RCS Patch Antenna," IEEE Antennas and Wireless Propagation Letters, vol. 13, pp. 1473–1476, 2014.
[63] K. Li, Y. Liu, Y. Jia, and Y. J. Guo, "A Circularly Polarized High-Gain Antenna with Low RCS over a Wideband Using Chessboard Polarization Conversion Metasurfaces," *IEEE Transactions on Antennas and Propagation*, vol. 65, no. 8, pp. 4288–4292, 2017.
[64] D. R. Smith, S. Schultz, P. Markoš, and C. M. Soukoulis, "Determination of Effective Permittivity and Permeability of Metamaterials from Reflection

and Transmission Coefficients," *Physical Review B*, vol. 65, p. 195104, April 19, 2002.

[65] C. L. Holloway, E. F. Kuester, J. Baker-Jarvis, and P. Kabos, "A Double Negative (DNG) Composite Medium Composed of Magnetodielectric Spherical Particles Embedded in a Matrix," *IEEE Transactions on Antennas and Propagation*, vol. 51, pp. 2596–2603, 2003.

[66] X. Chen, T. M. Grzegorczyk, B. I. Wu, J. Pacheco, Jr., and J. A. Kong, "Robust Method to Retrieve the Constitutive Effective Parameters of Metamaterials," *Physical Review E*, vol. 70, p. 016608, July 26, 2004.

[67] D. Samantaray, S. K. Ghosh, and S. Bhattacharyya, "Modified Slotted Patch Antenna with Metasurface as Superstrate for Dual-Band Applications," IEEE Antennas and Wireless Propagation Letters, vol. 22, no. 1, pp. 109–113, January 2023.

[68] F. Costa, S. Genovesi, and A. Monorchio, "On the Bandwidth of High Impedance Frequency Selective Surfaces," *IEEE Antennas and Wireless Propagation Letters,* vol. 8, pp. 1341–1344, 2009.

[69] Y. F. Cheng, J. Feng, C. Liao, and X. Ding, "Analysis and Design of Wideband Low-RCS Wide-Scan Phased Array with AMC Ground," IEEE Antennas and Wireless Propagation Letters, vol. 20, no. 2, pp. 209–213, February 2021.

[70] E. B. Lima, J. R. Costa, and C. A. Fernandes, "Multiple-Beam Focal-Plane Dual-Band Fabry–Pérot Cavity Antenna with Reduced Beam Degradation," *IEEE Transactions on Antennas and Propagation*, vol. 67, no. 7, pp. 4348–4356, July 2019.

[71] P. Xie, G. Wang, H. Li, Y. Wang, and X. Gao, "Reconfigurable Fabry-Pérot Cavity Antenna Basing on Phase Controllable Metasurfaces," *Advanced Radio Frequency Antennas for Modern Communication and Medical Systems*, IntechOpen, UK, chapter 72936, September 2020.

[72] H. Yang and X. Liu, "Wearable Dual-Band and Dual-Polarized Textile Antenna for On- and Off-Body Communications," *IEEE Antennas and Wireless Propagation Letters*, vol. 19, no. 12, pp. 2324–2328, December 2020.

[73] H. Zhai, K. Zhang, S. Yang, and D. Feng, "A Low-Profile Dual-Band Dual-Polarized Antenna with an AMC Surface for WLAN Applications," IEEE Antennas and Wireless Propagation Letters, vol. 16, pp. 2692–2695, 2017.

[74] J. Wu, S. Yang, Y. Chen, S. Qu, and Z. Nie, "A Low Profile Dual-Polarized Wideband Omnidirectional Antenna Based on AMC Reflector," IEEE Transactions on Antennas and Propagation, vol. 65, no. 1, pp. 368–374, January 2017.

[75] C. Chen, Z. Liu, H. Wang, and Y. Guo, "Metamaterial-Inspired Self-Polarizing Dual-Band Dual-Orthogonal Circularly Polarized Fabry–Pérot Resonator Antennas," IEEE Transactions on Antennas and Propagation, vol. 67, no. 2, pp. 1329–1334, February 2019.

[76] M. Mattsson, C. I. Kolitsidas, and B. L. G. Jonsson, "Dual-Band Dual-Polarized Full-Wave Rectenna Based on Differential Field Sampling," *IEEE Antennas and Wireless Propagation Letters*, vol. 17, no. 6, pp. 956–959, June 2018.

[77] F. Liu, J. Guo, L. Zhao, G. Huang, Y. Li, and Y. Yin, "Dual-Band Metasurface-Based Decoupling Method for Two Closely Packed Dual-Band Antennas," IEEE Transactions on Antennas and Propagation, vol. 68, no. 1, pp. 552–557, January 2020.
[78] S. B. T. Wang, A. M. Niknejad, and R. W. Brodersen, "Circuit Modelling Methodology for UWB Omnidirectional Small Antennas," *IEEE Journal on Selected Areas in Communications*, vol. 24, no. 4, pp. 871–877, 2006.
[79] D. Samantaray and S. Bhattacharyya, "A Metasurface-Backed Planar Low-Profile Dual-Band Monopole Antenna," *Journal of Electromagnetic Waves and Applications*, vol. 37, no. 7–9, pp. 884–897, May 2023.
[80] A. Ahmed, M. R. Robel, and W. S. T. Rowe, "Dual-Band Two-Sided Beam Generation Utilizing an EBG-Based Periodically Modulated Metasurface," *IEEE Transactions on Antennas and Propagation*, vol. 68, no. 4, pp. 3307–3312, April 2020.
[81] L. Gu, W. Yang, Q. Xue, and W. Che, "A Dual-Band Steerable Dual-Beam Metasurface Antenna Based on Common Feeding Network," IEEE Transactions on Antennas and Propagation, vol. 69, no. 10, pp. 6340–6350, October 2021.
[82] E. B. Lima, J. R. Costa, and C. A. Fernandes, "Multiple-Beam Focal-Plane Dual-Band Fabry–Pérot Cavity Antenna with Reduced Beam Degradation," *IEEE Transactions on Antennas and Propagation*, vol. 67, no. 7, pp. 4348–4356, July 2019.
[83] H. Zhai, K. Zhang, S. Yang, and D. Feng, "A Low-Profile Dual-Band Dual-Polarized Antenna with an AMC Surface for WLAN Applications," IEEE Antennas and Wireless Propagation Letters, vol. 16, pp. 2692–2695, 2017.

Chapter 3

Wideband and multi-band circularly polarized planar slot antennas

Kapil Saraswat and Trivesh Kumar

3.1 INTRODUCTION

An antenna is an important element of a wireless radio frequency (RF) communication system. The parameters that are used to describe an antenna could be broadly classified into input parameters and radiation parameters. The input parameter, e.g. input reflection coefficient and input impedance, etc. will indicate how well the antennas are matched to the port of the communication system. The far-field radiation parameters, such as polarization, radiation pattern, gain, efficiency, etc. indicate how the antenna radiates or receives RF signals [1].

In a wireless communication system, it is important to align the two transmitter and receiver antennas. One type of alignment is to ensure the direction of maximum radiation of both antennas coincides with each other. However, this is not sufficient to establish communication between the transmitter and receiver. It is also important to ensure that the fields of the two antennas are oriented properly.

The direction of the electric field vector in the far field of an antenna is used to describe one of its properties called 'polarization'. The polarization could be classified into three categories, viz., linear, circular, and elliptic, based on the shape of the locus. For efficient communication, the polarization of the transmit antenna should match the polarization of the receive antenna. For example, if the antenna is oriented such that it radiates vertically (with respect to ground) polarized waves, the receiving antenna should also be vertically polarized.

If the polarizations of the two antennas are not identical, the received power will be lower than that would have been obtained with the two identical or matched polarizations. This misalignment in polarization can be quantified using polarization mismatch. Consider a communication system consisting of linearly polarized transmit and the receive antennas, that are polarization matched. Under free space conditions, the polarizations are matched. However, due to environmental factors, such as scattering, shadowing, or reflections, the polarization of the transmitted wave can get

 DOI: 10.1201/9781003470854-3

changed by the time it reaches the receiver and hence may result in reduced received signal strength. In fact, if the receive antenna is vertically polarized and the incoming electromagnetic (EM) wave is horizontally polarized, the receiver does not get any signal at all.

The circularly polarized (CP) antennas which have orthogonal components of the electric field that have equal amplitude ±90° phase difference are widely being used in different wireless communication applications. CP antennas exhibit the following benefits over linearly polarized antennas,

a. Mitigation of multipath propagation or fading
b. Polarization losses due to misalignment
c. Immunity to Faraday rotation

The antennas presented in this chapter could have many applications such as:

a. When the antenna and telemetry systems are placed on a movable object, the proper antenna orientation cannot be guaranteed, and therefore, in such applications the CP antenna plays an important role, these applications are satellite navigation signals (GNSS), GPS, remote telemetry, etc.
b. In radio frequency identification (RFID) the CP pattern ensures RFID tag detection regardless of its orientation. This makes the RFID system more applicable in such areas including industries, shopping malls, employment management systems, etc.
c. In space, the satellites are using spin stabilization to maintain their orbit and path. As the satellite rotates continuously, the CP antennas are useful for satellite–Earth communication, which provides reliable communication despite the effect of Faraday rotation in the ionosphere and satellite spinning.
d. To improve the system capacity and efficiently handle multipath fading effects CP antennas are used.

This chapter outlines the conventional methods for the design of CP antennas. The measurement techniques and experimental setup for the different parameters of CP antennas have also been discussed.

3.2 CIRCULARLY POLARIZED ANTENNAS

Historically, before the era of microstrip antennas, widely used antennas generating circular polarization were the 'Turnstile' antenna [2] (also known as the crossed-dipole antenna), helix antenna [3], and a slot-dipole antenna configuration [4]. Figure 3.1 shows the structure of these antennas. As shown in Figure 3.1(a), the turnstile antenna consists of two orthogonally

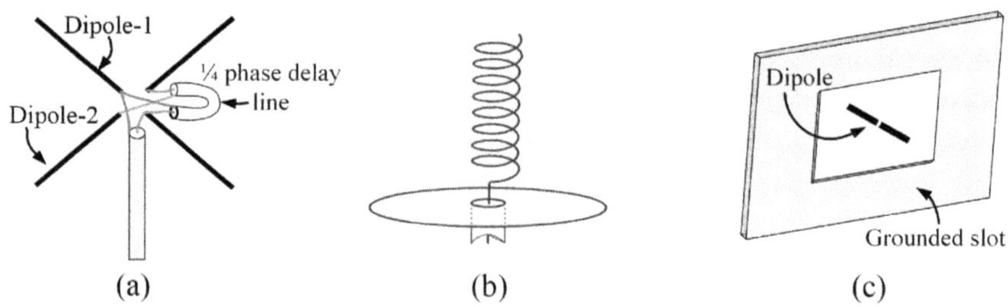

Figure 3.1 Circularly polarized antennas (a) Turnstile antenna, (b) helix antenna, and (c) slot-dipole antenna.

placed dipole antennas of equal length and a phase delay line. The Dipole-1 is directly connected to the coaxial line, and the Dipole-2 is connected to the coaxial feed line through a $\frac{\lambda}{4}$ phase line (provides 90° phase difference). Figure 3.1(b) shows a helix antenna. The helix can be described as a *monofilar* (one-wire) helical antenna [5]. Figure 3.1(c) shows the slot-dipole antenna configuration [4], where a tilted $\frac{\lambda}{2}$ dipole excites a grounded slot, which produces orthogonal field components resulting in circular polarization. Circular polarization can also be generated with circular or square aperture, but a complex feed network consisting of a 90° phase shifter is required. Microstrip patch antennas are planar structures and widely used antennas in recent days. Microstrip antennas having various patch shapes can radiate circular polarization. Figures 3.2-3.3 show different configurations including a nearly square patch, elliptical patch, square patch with 45° slot, square patch with perturbing slot, truncated corner square patch, etc. A single patch can generate CP when two orthogonal modes (or patch mode) are excited with equal magnitude and 90° phase difference. This can be accomplished in two different ways, by using a dual orthogonal fed patch antenna or by using a single fed patch antenna.

Figure 3.3 shows the single feed microstrip arrangement [7–9], where the patch antenna is excited by a single feed located in such a way that it excites two equal amplitude degenerate orthogonal modes. By introducing an asymmetry in the patch using a stub or a slot, the degeneracy of the modes can be removed. The perturbation will detune (depending on the location of the perturbing element) the frequency response of one of the modes in such a way that they have the same amplitude and are 90° out of phase with each other.

With the development of the CP microstrip antennas, researchers are designing many different types of microstrip-based antennas that can radiate circular polarization by incorporating symmetric and asymmetric

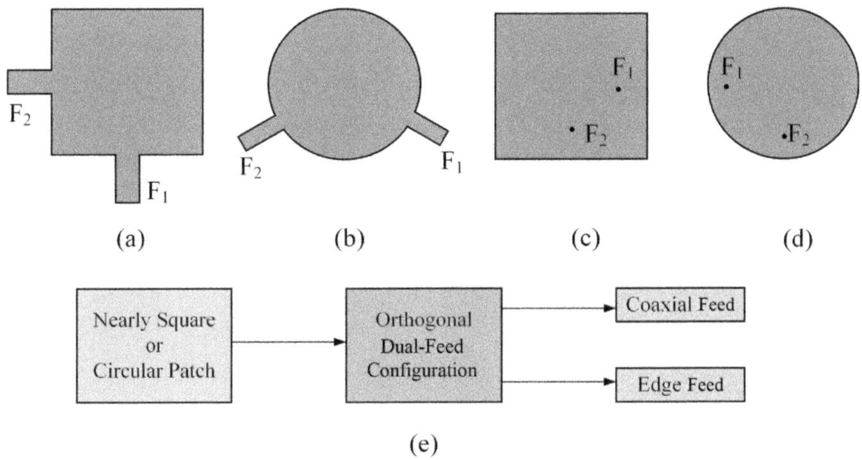

Figure 3.2 Dual orthogonal feed circularly polarized microstrip patch antennas (a) edge fed square patch, (b) edge fed circular patch, (c) direct (coaxial) fed square patch, (d) direct fed circular patch (F represent feed location), and (e) Patch antenna with dual spatially orthogonal feed arrangement. Figure 3.2 shows the microstrip antenna having dual spatially orthogonal feed arrangement [6, 7]. To generate CP, the two orthogonal feeds are excited with a 90° phase difference between them. This can be achieved by employing an external circuitry (phase-shifter, power divider, 90° hybrid, ring hybrid, etc.). To reduce the overall footprint of the antenna, the phase shifter (or other arrangements) can also be placed on the microstrip patch itself, and the patch can be excited by edge feed or inset fed, while the phase shifter is excited by a coaxial feed.

slots within the patch or at the corner of the patch. The major drawback of these microstrip antennas is the narrow CP bandwidth. Other conventional microstrip and planar configurations used for the CP radiation are the triangular patch antenna [10], E-shaped patch antenna [11], asymmetrically placed monopole antenna, and crossed dipole antenna, as shown in Figure 3.4. Similar to the conventional patch antenna, two orthogonal field components having 90° phase difference are produced in these structures which result in the circular polarization in the broadside direction.

Nasimuddin *et al.* show that CP radiation can be obtained by introducing asymmetric slots at the corners of a single feed patch antenna. Figure 3.5 shows the slot may be square [12], circular [13], any arbitrary slot [14], or asymmetric slits [15] loading the square patch. By loading the asymmetric slots or slits, fundamental resonant modes of the square patch are split into two near-degenerate resonant modes. The CP can be generated by selectively tuning the slot dimensions and their position on the patch antenna. Therefore, other possible planar antenna configurations were investigated by the researchers for the broadband CP response.

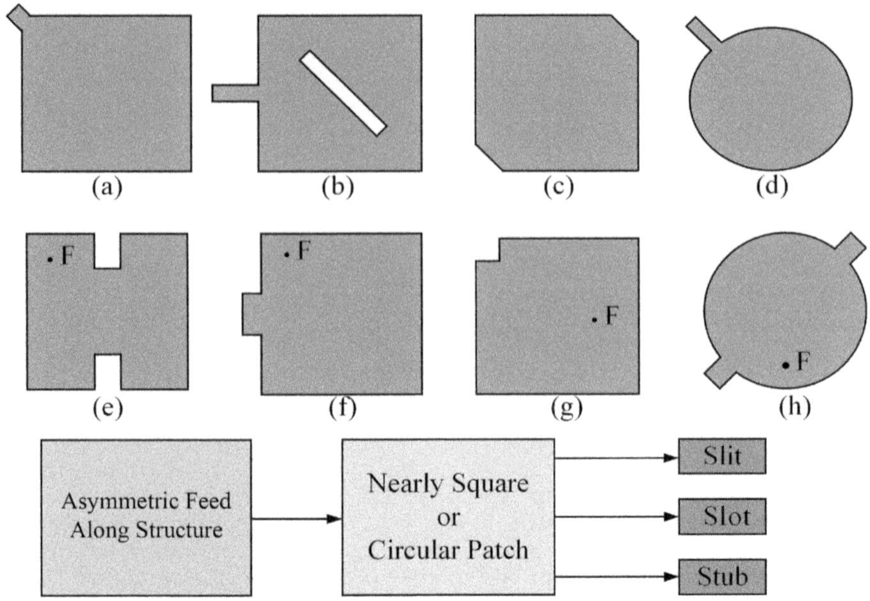

Figure 3.3 Single feed circularly polarized microstrip patch antennas (a) almost square, (b) patch with 45° slot, (c) truncated corner patch, (d) elliptical patch (e) edge slot perturbation, (f) edge stub perturbation, (g) corner perturbation, and (h) circular patch with stub perturbation (F represent feed location).

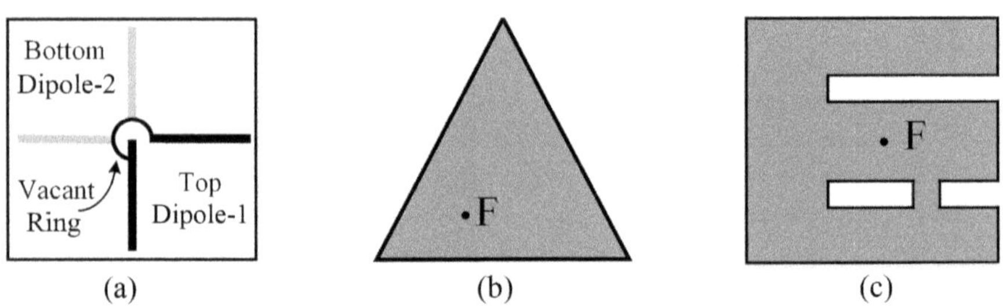

Figure 3.4 Circularly polarized antennas (a) printed crossed dipole, (b) triangular patch, and (c) asymmetric E-shaped patch.

A monopole antenna can generate circular polarization if an asymmetrical structure is used to produce two orthogonal field components. It can be obtained by various monopole designs shown in Figure 3.6. For example, as shown in Figure 3.6, a monopole is placed with an asymmetric ground plane [16] and in [17] a monopole structure is excited by an offset feed as shown in Figure 3.6. Similarly, asymmetry in the monopole can be introduced by adding stubs to the radiating element, by adding stubs in the

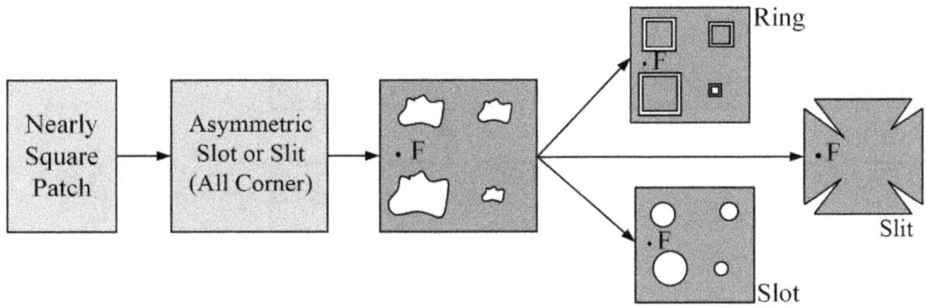

Figure 3.5 Circularly polarized microstrip patch antennas [12–15] using arbitrary shaped slot (or slit) loaded at the corner of microstrip patch antenna.

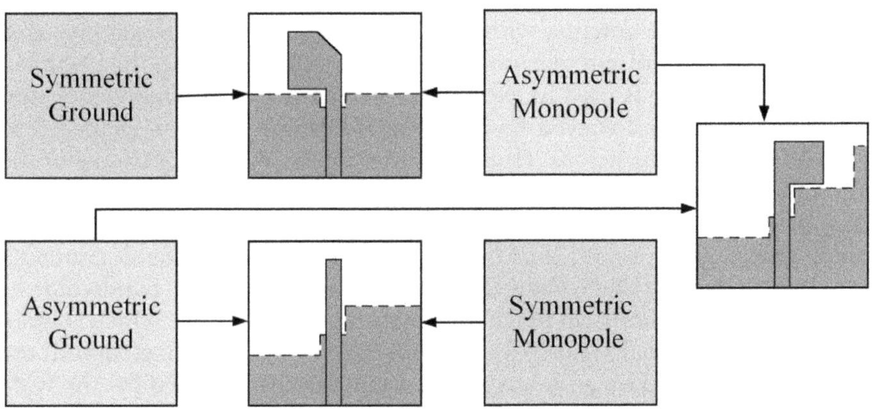

Figure 3.6 Circularly polarized monopole antennas using asymmetric ground or asymmetric fed monopole or their combination.

Source: Reference 16, reproduced courtesy of The Electromagnetics Academy.

ground plane, or by embedding slots in the ground plane. For example, in references [18–21], asymmetric monopole (such as falcate-shaped, fractal-shaped, or monopole with multiple arms) is used to excite orthogonal field components and additional grounded stubs or slots embedded to achieve equal magnitude and 90° phase difference for the generation of circular polarization.

In the recent past, to achieve wideband circular polarization, planar slot antennas have been used. These antennas are easy to fabricate and have a simple geometry. Some of these antennas have wide bandwidth, low profile, lightweight, etc. Figure 3.7 shows the basic square, circular slot, and arbitrarily shaped slots excited by two orthogonal feeds. Equal power is fed to the two ports using the Wilkinson power divider, and with the help

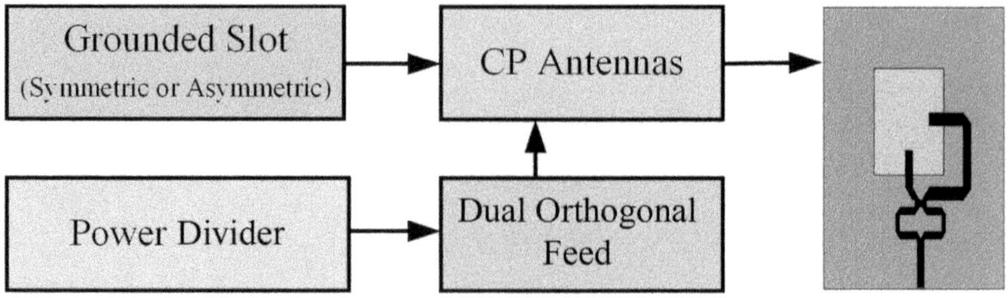

Figure 3.7 Circularly polarized slot antennas [22–28] using orthogonal fed slot antenna.

of an additional λ/4 transmission line, the phase difference is maintained. To reduce the complexity associated with the dual orthogonal feed arrangement, a slot antenna with a single microstrip feed can also be used for the generation of circular polarization. In reference [28], an L-shaped microstrip feed line is used to excite two orthogonal field components, and the phase of 90° is achieved by choosing the length of the L-shaped feed line and the slot dimensions. Due to the simplicity, different variants of the slot antenna excited by an L-shaped feed line were proposed [22–27]. For example, annular ring-shaped slots [22–24], square slots [25], and another circular ring-inspired slot [26, 27] were studied in the past. The slot antenna [22–25] excited by the L-shaped fed line enhances the CP bandwidth as compared to the microstrip patch antenna radiating CP wave which is more than 11% (3 dB axial ratio (AR)). A parasitic ring [28] is coupled with the slot antenna excited by an L-shaped feed line can also be used for the wide CP bandwidth. The CP bandwidth was further improved (more than 85%) by a circular ring-inspired slot [27] by introducing a resistor at the end of the microstrip feed line, which is shorted to ground, to absorb the power at the cost of significantly lower efficiency.

The impedance bandwidth and AR bandwidth of a slot antenna fed by a single transmission line can be enhanced by merging multiple modes. This can be achieved by exciting linked slots etched in a ground plane by a microstrip feed as shown in Figure 3.8. In this configuration, two bands are obtained by exciting the primary slot and linked slot. By keeping both the bands close to each other, it is possible to generate wideband CP. For example, the square [29, 30] and circular [31] shaped linked slots can radiate wide circular polarization. To improve the CP bandwidth, elliptical linked slots were used along with two rectangular stubs embedded in the feed line [32]. The location of slots, the distance between them, and the slot dimensions are used to optimize the CP bandwidth.

For wider CP bandwidth, substantial work has been done on CP slot antennas. Several single-feed slot-based antennas have been studied, and

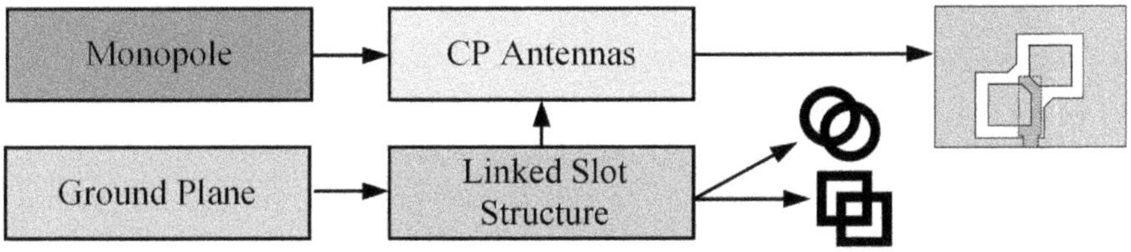

Figure 3.8 Circularly polarized grounded linked slot antennas [29–31].

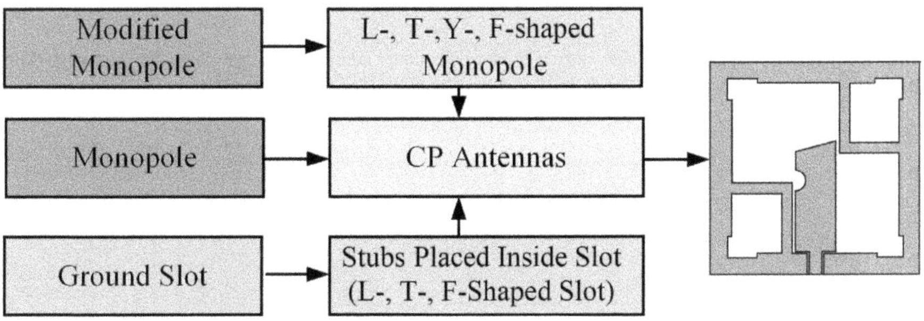

Figure 3.9 Circularly polarized grounded slot antennas with modified feed and stubs/slots inside the grounded slot.

various techniques have been adopted for the generation of wideband circular polarization. The coplanar waveguide (CPW) fed slot antennas have several advantages over the microstrip fed slot antennas, such as wider bandwidth, low dispersion, simple realization due to single layer arrangement, etc. The CPW facilitates the integration of lumped elements on the same side as the antenna. The CPW-fed slot structure can be made to radiate CP waves by embedding appropriate features as shown in Figure 3.9 to redistribute the magnetic currents such that the slot supports two orthogonal modes having equal amplitude and 90° phase difference. Conventionally, a CPW-fed square ring slot monopole antenna excites TM_{11} mode which results in a linearly polarized wave. However, when the perturbation is introduced within the slot, two orthogonal modes, viz, TM_{01} and TM_{10} are excited, which results in the CP radiation. For example, a broadband circular polarization is achieved by introducing a T-shaped stub toward the center of a slot which is fed by a CPW-fed monopole [33]. In the other example, as depicted in Figure 3.9, a pair of inverted L-shaped stubs are placed inside the slot [34, 35] and a modified monopole is used to excite the slot, which results in the wideband CP radiation. Similarly, in reference [36] a pair of spur-shaped slits having unequal length is placed in the ground plane of a circular slot for the generation of circular polarization.

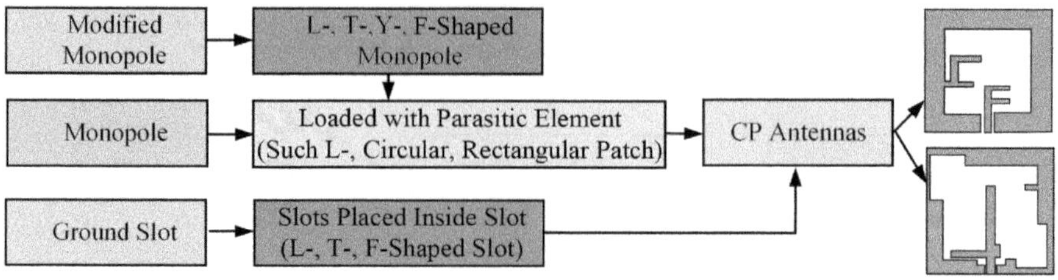

Figure 3.10 CPW fed circularly polarized slot antennas excited with modified feeding structure and slot/stubs inside the ground slot with parasitic elements.

Source: References 39–40, reproduced courtesy of The Electromagnetics Academy.

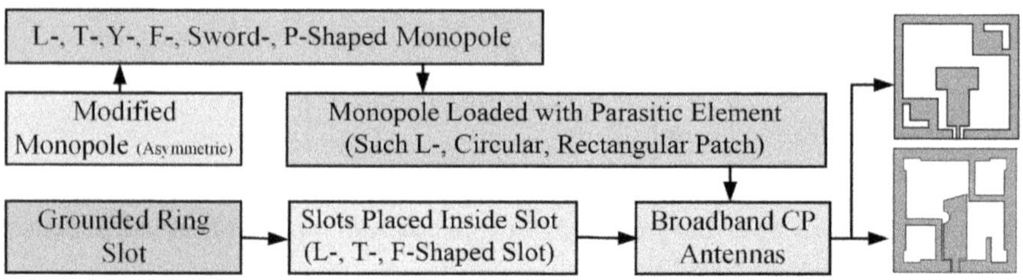

Figure 3.11 Broadband circularly polarized slot antennas using modified monopole, stubs/slots inside ground plane with additional grounded/non-grounded parasitic element/stub.

Source: Reference 43, reproduced courtesy ofThe Electromagnetics Academy.

Figure 3.10 illustrates different CPW-fed slot antenna configurations utilizing various perturbations [37–42] for the generation of wideband circular polarization.

A grounded slot excited by an asymmetrically placed monopole antenna has been used to enhance the CP bandwidth, and a few arrangements are shown in Figure 3.11. As depicted in Figure 3.11, a monopole is placed asymmetrically within the grounded ring which is loaded with a grounded stub [43, 44]. The coupling of the monopole with the perturbing element and the edge of the grounded ring excites the orthogonal fields, and the spacing between the ring and the monopole gives the required phase difference to produce a CP wave. Similarly, a Y-shaped monopole [45] having unequal arms is placed within the square-shaped grounded ring which radiates the CP wave. In another example, an L-shaped monopole is asymmetrically placed in a grounded ring [46] and the arm of the L-shaped monopole coupled with the horizontal and vertical side edge of the grounded ring

Table 3.1 Comparison of CP antennas with proposed antenna

Ref.	*Size* mm^2	*Frequency range* $S_{11} < -10$ *dB*	*(% Bandwidth)* *AR < 3 dB*
[36]	20 × 20	27.27%	12.72%
[37]	70 × 70	39.6%	12.4%
[46]	60 × 60	5.2%	3.8%
[50]	50 × 50	63.89%	7.93%
[51]	55 × 55	80%	8%
[52]	90 × 16.2	4.3%	4.3%
[53]	300 × 300	2.15%	2.15%
[54]	45 × 45	25.66%	10.56%
[47]	33 × 50	35.16%	26.21%
[48]	40 × 18.5	29.45%	12.75%
[49]	23.5 × 22	43.62%	30.5%

$\lambda_{L,AR}$: wavelength corresponding to the lower band edge frequency where the AR is 3dB.

results in the wideband CP radiation. Similarly, discontinuous ring slot can also radiate circular polarization [47–49]. Table 3.1 shows the comparison of CP antenna with the proposed antenna.

3.3 TECHNIQUES FOR CP BANDWIDTH ENHANCEMENT

Various techniques have been proposed by the researchers to enhance the bandwidth of the CP antennas, such as deformation of the ground plane, introduction of stubs and slots, use of parasitic elements, stacked patch configurations, and multiple radiators with sequential phase feed, etc.

The parasitic elements, such as a parasitic ring, patch, or stub are placed within the slot or close to the radiator to excite the higher-order modes or tune the electric field components to achieve wideband CP response. Figure 3.12 shows different antenna structures and methods where parasitic elements are used to achieve wideband CP radiation. In reference [55], a C-shaped monopole is asymmetrically placed on a ground plane with a vertical stub, resulting in CP radiation. When an open loop is coupled with the monopole, it induces a current similar to a traveling-wave mode and contributes to wideband CP radiation [55]. Similarly, in reference [56] a pair of parasitic curl-shaped strips is juxtaposed to the monopole. The asymmetric parasitic strips generate two orthogonal fields, and multi-resonant modes are generated by multi-resonant paths (due to the spiral geometry of the parasitic loop) which results in wideband circular polarization.

An asymmetrically fed L-shaped monopole antenna placed inside a slot antenna radiates circular polarization. To enhance the bandwidth, a parasitic patch is introduced within the slot. For example, in reference [57],

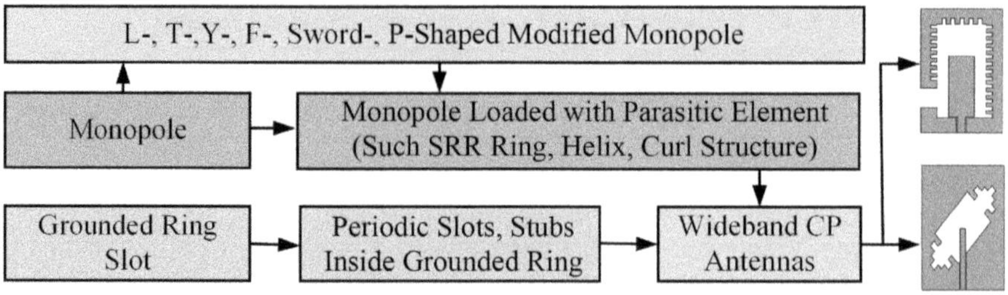

Figure 3.12 Wideband circularly polarized antennas using periodic slots/stubs and parasitic elements to obtain multimode operations to achieve wide bandwidth.

a CPW-fed grounded ring antenna structure consisting of an L-shaped monopole placed within the grounded ring along with a horizontal stub for broadband CP radiation. A rectangular parasitic patch is placed close to the L-shaped monopole for bandwidth enhancement. With similar configurations, different parasitic patches, such as a square [58, 59], vertical strip [60], etc. can be used for the bandwidth enhancement. A circular [61] or square [62] shaped parasitic patch can be placed on the top of a circular slot antenna which is excited by an L-shaped feed for the wideband CP operation.

Various slot antennas consisting of grounded stubs or slots in the grounded ring exhibit wide AR bandwidth. For example, in references [63, 64] perturbation slots are included within a truncated corner grounded circular slot to build a higher order (Quasi-TE31) mode. In reference [65], an L-shaped radiating patch is placed inside the ground plane and to obtain wide AR bandwidth one end of the L-shaped patch is shorted to the ground plane via a slit. Nagendra *et al.* [65] also use the same techniques in a CPW-fed slot antenna for wide AR bandwidth. A pair of hat-shaped patches [66] is placed at 45°, and 135° within the annular ring slot antenna which is excited by an L-shaped monopole for wider AR bandwidth, and impedance matching is improved by deforming the monopole.

Another popular technique to enhance AR bandwidth is the rotated radiating structure excited by sequential phase feed. In 1989, P. S. Hall [69] showed that a sequentially rotated feeding of CP microstrip antenna array gives a substantial improvement in the antenna performance such as impedance bandwidth and AR bandwidth. In that work, four CP microstrip patch antennas are arranged in a rotationally symmetric structure and excited by a sequentially rotated feeding network. All four antennas are simultaneously excited with a 90° phase difference between each element. The feed system can achieve wide bandwidth. However, the performance of the overall antenna is determined by the bandwidth of the radiator [69–71]. Therefore, based on this arrangement, an array of simultaneously excited

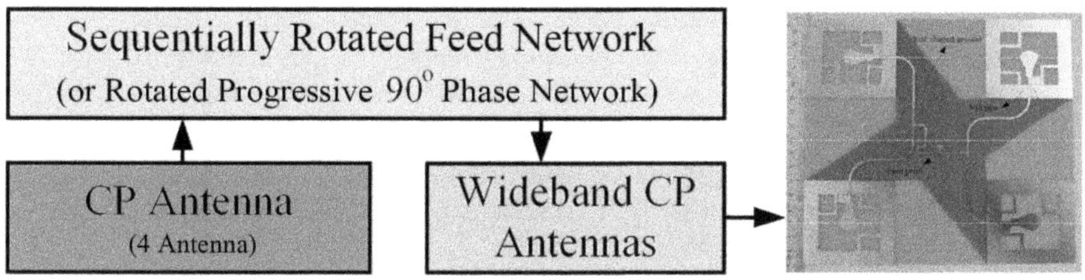

Figure 3.13 Sequentially rotated circularly polarized antennas.

Source: Reference 68, reproduced courtesy of The Electromagnetics Academy.

four-slot antenna elements producing CP wave is proposed for CP bandwidth enhancement [67, 68, 72], and illustrated in Figure 3.13.

To realize sequentially rotated CP antenna arrangement, various techniques are used. As shown in Figure 3.13, an annular antenna is excited by the orthogonal feed network through grounded slits which act as a CP-radiating antenna element [67, 68, 72]. This annular ring arrangement is sequentially rotated by 90° and excited by the sequentially rotated ring-shaped microstrip network which results in a considerably wide circular polarization. Further, it is sequentially rotated and excited by the power divider arrangement consisting of three T-junctions to divide the power into four ports. The phase difference to the sequentially rotated CP radiator is provided by extending the length of the power divider arm. These antennas give very wide CP bandwidth, but the major drawback of such techniques is the size of the antenna (consisting of four radiating elements, and their spacing) and the complexity associated with the feed network. Table 3.2 shows different wideband CP antennas.

Square grounded slot antenna with modified feed-line such as Y-shaped branch [77], halberd shaped feed [78], spiraled monopole feed [79], using pair of inverted L-shaped strip [80], lightening-shaped feed-line [73, 81], meandered grounded ring, etc. also radiate wideband circular polarization. In reference [74], a grounded vertical stub is placed near to monopole to enhance the AR by utilizing characteristic mode analysis. A corrugated G-shaped grounded discontinuous ring slot antenna [75] is used for multiband circular polarization and a detailed mathematical analysis of the antenna using thin wire approximation is also presented to gain the physical insight for the CP radiation. In reference [76] a fractal-inspired tilted slot antenna is used for wide circular polarization.

3.4 TECHNIQUES FOR MULTI-BAND CP RADIATION

Similar to wideband antennas, the multi-band CP antennas are essential to multi-radio systems. This section reviews different techniques that utilize

Table 3.2 Different wideband CP antennas

Ref.	*Size* $\lambda^2_{L,AR}$	*Frequency range* $S_{11} < -10$ dB	*(% Bandwidth)* AR < 3 dB
[57]	0.38 × 0.38	72.6%	39.3%
[67]	0.42 × 0.38	88%	64.7%
[72]	1.13 × 1.13	33.5%	24.5%
[70]	1.35 × 1.35	20.8%	17.6%
[71]	1.2 × 1.2	54.3%	42.14%
[58]	0.44 × 0.37	63.8%	48.2%
[77]	0.31 × 0.31	43.4%	43.4%
[78]	0.42 × 0.42	49.6%	25.7%
[79]	0.36 × 0.36	45.1%	21.3%
[80]	0.46 × 0.46	62.3%	27.3%
[73]	0.41 × 0.41	51.4%	48.9%
[81]	0.57 × 0.57	135.6%	46.4%
[82]	0.31 × 0.31	63.1%	36.8%
[74]	0.37 × 0.	64.66%	64.66%
[75]	0.22 × 0.36	47.9 %	32.71%
[76]	0.61 × 0.78	22.5%; 63.5%	34.7%

$\lambda_{L,AR}$: wavelength corresponding to the lower band edge frequency where the AR is 3dB.

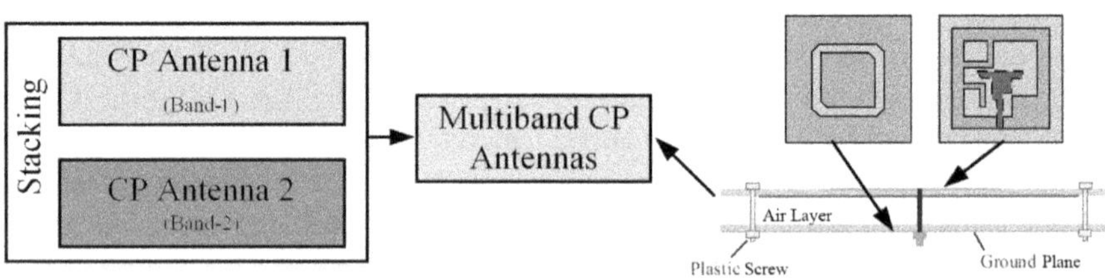

Figure 3.14 Dual-band circularly polarized antennas using single feed stacked microstrip patch antenna.

Source: Reference 83, reproduced courtesy of The Electromagnetics Academy.

stacked patches, multi-layer elements, stubs, slots, etc. to realize a multi-band CP antenna.

Two single-band elements can be stacked one above the other to realize a dual-band CP antenna [83–85]. Figure 3.14 shows the stacked CP antennas for dual-band operation. As shown in Figure 3.14, one truncated corner patch antenna and one wide-band CP slot antenna are stacked [83] for dual-band dual-polarization application, where both the stacked antennas are CP. Both the bands can be made CP by replacing the lower patch with a truncated corner microstrip patch. Similarly, in reference [84], two slot ring radiators are stacked and excited by a coupled line for dual-band

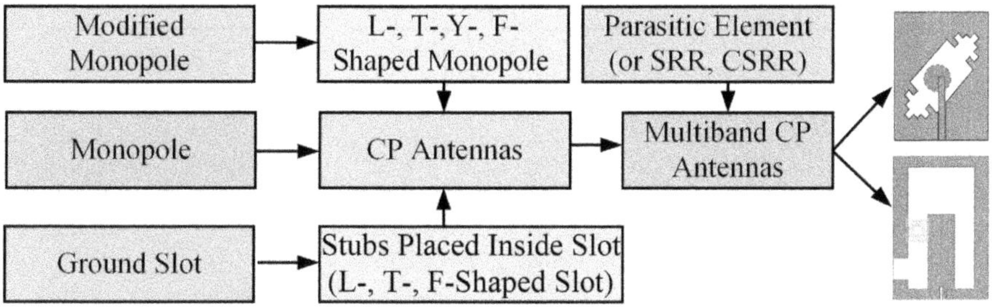

Figure 3.15 Dual-band circularly polarized slot antennas using parasitic patch or split ring (SR) loaded monopole placed grounded slot [108, 109].

dual-circular polarization. In references [86, 87], two eccentric rings are used in the microstrip configuration, and an arc-shaped conducting strip is used to excite dual CP by exciting two near-degenerate modes by splitting the corresponding TM11 mode of each of the eccentric rings. The limitations of these configurations are the size and the narrow CP bandwidth.

The grounded slot antenna can produce dual-/multi-band CP wave by incorporating a multi-arm monopole, antenna loaded with parasitic element, linked asymmetric slot, or using split ring resonators. Figure 3.15 shows different antenna structures radiating dual-band circular polarization. As shown in Figure 3.15, a discontinuous grounded ring slot antenna is excited by an asymmetrically placed SRR. In reference [88], a rectangular antenna is excited by a U-strip. The asymmetric arm length of the monopole results in two different resonant frequencies. The length of the U-strip arm decides the resonant frequency [88]. When a parasitic element is placed close to the CP structure, higher order modes can be generated, which results in multi-band CP radiation [89–92]. For example, an L-shaped monopole placed in a slot is coupled by a J-shaped parasitic strip [89] and used for the dual-band application. In reference [91], a cross-slot loaded circular patch placed inside a circular slot consisting of two slits is used to radiate CP waves at two different frequencies. Table 3.3 shows different multiband CP antennas.

The stubs and slots within the grounded ring or feed line (which excite the grounded slot) are also widely used for multi-band CP radiation [93, 95–97]. In references [93, 95], two rectangular strips are placed diagonally within a rectangular slot and excited by an L-shaped microstrip feed for multi-band CP radiation. Two asymmetric connected slots [93] fed by an L-shaped monopole, result in dual CP, and by integrating horizontal slots at the lower grounded strip, additional CP bands are produced.

The conventional CP antennas in concomitance with sub-wavelength resonators are also being used to realize the multi-band CP radiation [94, 98, 110, 111]. In reference [94], a pair of circular split ring resonators are

Table 3.3 Different multi-band CP antennas

Ref.	*Size* $\lambda^2_{L,AR}$	*Frequency range* $S_{11} < -10$ *dB*		*(% Bandwidth)* *AR < 3 dB*	
[83]	0.18 × 0.18	2.80%	9.00%	0%	2.40%
[84]	0.42 × 0.42	4.60%	3.70%	0.90%	0.40%
[85]	0.5 × 0.5	1.07%	1.19%	1.07%	1.19%
[87]	0.4 × 0.4	35%	14.60%	4.90%	4.60%
[88]	0.375 × 0.3	100%	7.20%	4.10%	
[91]	0.4 × 0.4	5.92%	4.29%	5.54%	2.46%
[89]	0.31 × 0.31	55.33%	20.98%	32.14%	20.98%
[86]	0.25 × 0.25	1.80%	2.60%	0.40%	0.60%
[93]	0.5 × 0.5	38.80%	27.50%	37.40%	16.30%
[94]	0.71 × 0.71	13.10%	13.60%	3.20%	4.20%
[98]	0.5 × 0.5	8.10%	28.10%	3.10%	4.20%
[99]	0.54 × 0.58	3.60%	3.30%	0.67%	0.60%
[100]	0.25 × 0.25	13.40%	0.70%	2.40%	
[101]	0.33 × 0.33	16.10%	12.40%	6.80%	0.60%
[102]	0.52 × 0.52	18.60%	18.40%	4.50%	3.50%
[103]	0.63 × 0.63	3.40%	6.90%	3.40%	8.06%
[104]	0.4 × 0.4	3.70%	1.20%	0.90%	0.60%
[105]	0.3 × 0.3	12.80%	7.60%	12.80%	7.60%
[106]	0.35 × 0.35	8.70%	23.00%	8.40%	19.20%
[107]	0.34 × 0.34	17.00%	20.70%	9.10%	11.10%
[108]	0.24 × 0.37	37.52%	4.80%	32.13%	2.17%
[109]	0.38 × 0.49	26%	56%	20.22%	31.50%

$\lambda_{L,AR}$: wavelength corresponding to the lower band edge frequency where the AR is 3dB.

placed above a monopole antenna exciting a truncated grounded slot. The truncated slot antenna radiates the CP wave, and the magnetic resonance of the SRR is used to produce another band without affecting the natural resonance of the slot antenna. A similar approach is used in references [110, 111] for dual-band CP radiation. In reference [98], multi-band LC resonators were embedded with the CP antenna for multi-band CP radiation. In reference [99], the author explores an L-shaped asymmetric strip connected at the edge of the radiator and in reference [100], a cross-dipole is loaded with a near-field resonant parasitic element for multi-band circular polarization. Two monofilar spiral slots are placed one inside another [102] and two slot coupled orthogonal feed lines [103] for dual-band circular polarization. Similarly, in reference [104] two grounded square ring slots excited by an L-shaped feed line are used for dual-band CP radiation. In references [105–107] perturbing element was used for the generation of multiband circular polarization. A CP antenna loaded with parasitic elements can also radiate circular polarization, such as in reference [108] a split ring resonator, and in reference [109] a circular disk acts as a parasitic element and radiates dual-band circular polarization.

3.5 RADIATION PATTERN MEASUREMENT

In general, two different methods are used to measure the CP radiation pattern. The first technique is simple and is used to measure the standard CP antenna (such as standard CP horn antenna or wideband helix antenna). Figure 3.16(a) shows the antenna range setup for the measurement of different far-field parameters inside an anechoic chamber. When AUT (antenna under test) is a CP, the standard antennas should also be CP of the same sense.

The second technique uses a standard linearly polarized directional antenna for the radiation pattern, and in the post-processing, the LP pattern is converted to the CP pattern. Figure 3.16, the CP AUT and a wideband reference horn antenna (standard horn) are positioned facing each other and are connected to a vector network analyzer (VNA). AUT is rotated about the vertical axis (θ rotation) from 0° to 360° using the stepper motor, M1, for horizontal and vertical polarization of the standard horn antenna is rotated (using ψ rotation). A software program controls the rotation of AUT and for each step, phase and magnitude are measured for both orthogonal polarization of standard horn. The left-hand and right-hand circular polarization wave components can be obtained as

$$E_{RHCP} = \frac{1}{\sqrt{2}}\left(E_H + jE_V\right) \tag{3.1}$$

$$E_{LHCP} = \frac{1}{\sqrt{2}}\left(E_H - jE_V\right) \tag{3.2}$$

where E_H and E_V are the vertical and horizontal field components and are given by

$$E_H = Re\left\{E_H\right\} + jIm\left\{E_H\right\} \tag{3.3}$$

$$= \left|E_H\right| e^{j\theta_H} = \left|E_H\right|\left[\cos\theta_H + j\sin\theta_H\right] \tag{3.4}$$

$$E_V = Re\left\{E_V\right\} + jIm\left\{E_V\right\} \tag{3.5}$$

$$= \left|E_V\right| e^{j\theta_V} = \left|E_V\right|\left[\cos\theta_V + j\sin\theta_V\right] \tag{3.6}$$

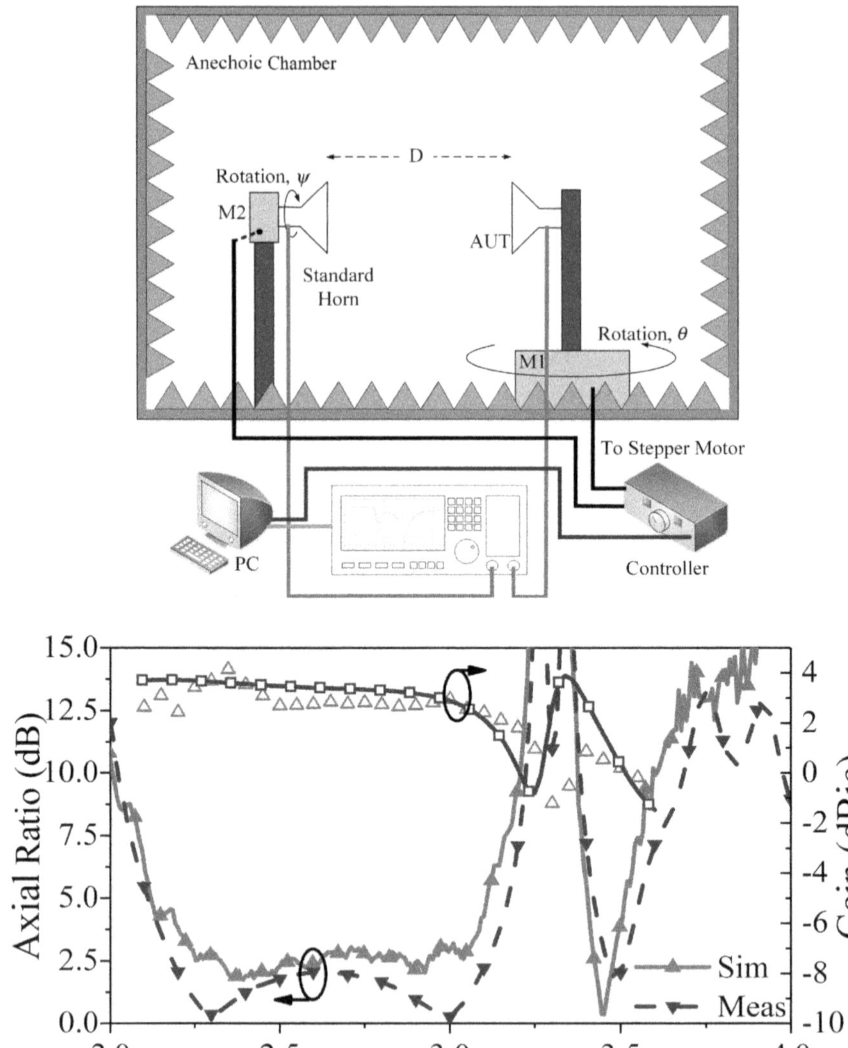

Figure 3.16 CP antenna measurement (a) experimental setup for the measurement of the radiation pattern, gain, and AR of circularly polarized antennas, (b) AR measurement in single direction (in this case broadside direction), and (c) measurement in θ-plane (data processed and plotted in Python).

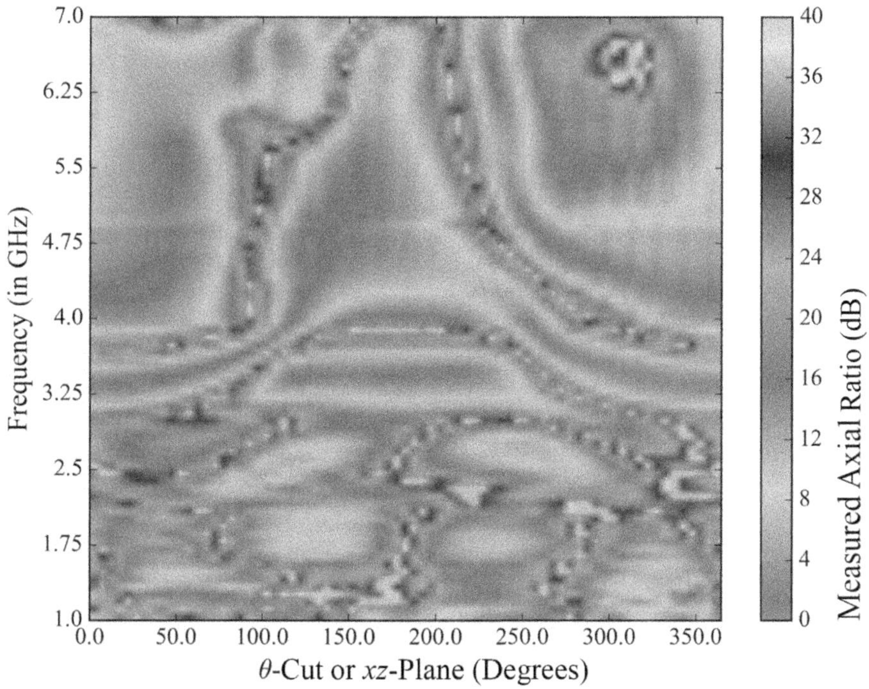

Figure 3.16 (Continued)

Substituting eq. (3.3) and (3.5) into (3.1) and (3.2) we get

$$E_{LHCP} = \frac{1}{\sqrt{2}}\left(|E_H|\cos\theta_H + j|E_H|\sin\theta_H\right) + \frac{j}{\sqrt{2}}\left(|E_V|\cos\theta_V + j|E_V|\sin\theta_V\right) \tag{3.7}$$

$$E_{RHCP} = \frac{1}{\sqrt{2}}\left(|E_H|\cos\theta_H + j|E_H|\sin\theta_H\right) - \frac{j}{\sqrt{2}}\left(|E_V|\cos\theta_V + j|E_V|\sin\theta_V\right) \tag{3.8}$$

The left-handed and right-handed CP radiated powers, P_{LHCP} and P_{RHCP} in dB are computed from eq. (3.7) and (3.8). which can be implemented and plotted (polar plot) using MATLAB® or Python programming. E_{RHCP} and E_{LHCP} patterns can be normalized and compared with the normalized simulated radiation pattern.

3.6 AXIAL RATIO MEASUREMENT

The purity of the CP wave is defined by the AR. The minor and major axes are the minimum (E_{min}) and maximum (E_{max}) of the measured electric field, respectively. The AR can be expressed by

$$AR(dB) = 20\log_{10}\left(\frac{E_{max}}{E_{min}}\right) \tag{3.9}$$

$$= 20\log_{10}\left[\frac{\left|E_{RHCP} + E_{LHCP}\right|}{\left|E_{RHCP} - E_{LHCP}\right|}\right] \tag{3.10}$$

There are two different methods used for the measurement of the AR. In the first method, P_{RHCP} and P_{LHCP} calculated in the previous step are used to calculate the AR using

$$AR(dB) = 20\log_{10}\left(\frac{1+10^{-P_{dB}/20}}{1-10^{-P_{dB}/20}}\right) \tag{3.11}$$

where $P_{dB} = |P_{RHCP} - P_{LHCP}|(dB)$ is the cross-polar level measured in the direction of the main beam. The major advantage of this method is that along with the CP radiation pattern; the AR can be obtained without extra measurement.

In another method, an AR plot is retrieved using spinning a linearly polarized antenna in the measurement. The standard horn antenna is rotated from 0° to 180° at a step of 10°, and for each step, the field is measured. The difference between the global maxima and the minima of the measured field is the AR for each frequency point. This method takes more time compared to the previous one. With these techniques, a line plot of the AR (AR v/s frequency) or 3D surface plot (AR v/s frequency v/s $_{\theta}$) can be plotted using MATLAB® or Python script. Figures 3.16(b) and 3.16(c) show the measured AR of two different antennae in two different ways. In Figure 3.16(b), AR is measured in a single direction and plotted. In Figure 3.16(c), AR is measured and plotted over the entire cut plane (θ- or φ- cut plane) over the band of frequencies.

3.7 GAIN MEASUREMENT

The measurement of the gain of CP antennas is done through the gain substitution method. Let, G_d be the difference between the measured power level of the standard horn antenna, $G_{std,}$ and the absolute gain of the AUT, G_o(dBil) or $G_o = G_{std} - G_d$. The gain of AUT (G_{AUT}) is then expressed as

$$G_{AUT}\,(dBic) = G_o + 3 \tag{3.12}$$

A 3-dB increase in the G_{AUT} is the difference in the power received by a linear isotropic radiator (LP antenna) and circular isotropic radiator (CP antenna) when measuring the received power of a polarization pure CP signal (AR = 0). A gain correction factor is incorporated due to the cross-polar power (for finite AR, i.e., AR > 0) for a practical CP wave, the CP antenna gain can be expressed as

$$G_{AUT}\left(dBic\right) = G_o + 3 + 20\log_{10}\left[\frac{1+10^{-\frac{AR}{20}}}{2}\right] \tag{3.13}$$

where G_o (dBil), can be calculated as, $G_o = G_{std} - \triangle$. The G_{std} is calculated from the friis formula. Equations (3.13) or (3.11) can be implemented and plotted (line plot) in MATLAB® or Python.

3.8 CONCLUSION

In this chapter, a review of different CP antenna systems has been presented. This chapter covers some initial work on CP antennas and planar slot antennas generating circular polarization. This chapter concludes that CP antennas are superior in terms of polarization-loss, Faraday rotation, susceptibility to multipath and fading, etc. It presents initial work on the planar CP antennas and different techniques for the generation of circular polarization.

To achieve bi-directional CP radiation, a slot antenna was used utilizing planar fabrication techniques and various techniques were discussed in dentils for the narrowband CP generation. Next, this chapter highlights some methods for the generation of wideband circular polarization. Techniques such as characteristic mode analysis, magnitude and phase of field components, surface current distribution, etc. have been used as an additional tool to improve the antenna performance. The chapter further discussed different methods to achieve multiband circular polarization by incorporating slits, metallic strips, parasitic elements, split-ring-resonator, etc. It is concluded that several methods achieve wideband and multiband circular polarization response which are basically either merging or splitting the band through perturbation of the field.

The measurement of the CP antenna parameter is also important, and therefore in this chapter, a concise discussion on the measurement of antenna parameters has been presented along with the experimental setup. Python or MATLAB® script can be developed using the discussed equations for in-house measurement systems and for post-processing of different antenna parameters.

REFERENCES

[1] C. A. Balanis. *Advanced Engineering Electromagnetics*. Wiley, 2nd edition, 2012.

[2] G. Brown. The turnstile. *Electronic*, 9:14–17, April 1936.

[3] J. D. Kraus. The helical antenna. *Proceedings of the IRE*, 37(3):263–272, March 1949.

[4] E.J. Wilkinson. A circularly polarized slot antenna. *Microwave Journal*, 4(3):97–100, March 1961.

[5] J. D. Kraus. The helical antenna. *Proceedings of the IRE*, 37(3):263–272, March 1949.

[6] P. Bhartia, I. Bahl, R. Garg, and A. Ittipiboon. *Microstrip Antenna Design Handbook*. Artech House, 2001.

[7] S. S. Gao, Q. Luo, and F. Zhu. *Circularly Polarized Antennas*. Wiley-IEEE Press, City, 2013.

[8] D. M. Pozar and D. H. Schaubert. *Microstrip Antennas: The Analysis and Design of Microstrip Antennas and Arrays*. IEEE, 1995.

[9] J. R. James and P. S. Hall, editors. *Handbook of Microstrip Antennas*, Volume 1. IET, January 1989.

[10] J. H. Lu, C.L. Tang, and K.L. Wong. Circular polarisation design of a single-feed equilateral-triangular microstrip antenna. *Electronics Letters*, 34(4):319, 1998.

[11] F. Yang, X. X. Zhang, X. Ye, and Y. Rahmat-Samii. Wide-band E-shaped patch antennas for wireless communications. *IEEE Transactions on Antennas and Propagation*, 49(7):1094–1100, July 2001.

[12] X. Q. Nasimuddin and Z. N. Chen. A compact circularly polarized slotted patch antenna for GNSS applications. *IEEE Transactions on Antennas and Propagation*, 62(12):6506–6509, December 2014.

[13] Z. N. C. Nasimuddin and X. Qing. Asymmetric-circular shaped slotted microstrip antennas for circular polarization and RFID applications. *IEEE Transactions on Antennas and Propagation*, 58(12):3821–3828, December 2010.

[14] Z. N. C. Nasimuddin and X. Qing. Compact circularly polarized asymmetric slotted microstrip patch antennas. *Microwave and Optical Technology Letters*, 54(8):1920–1927, May 2012.

[15] X. Q. Nasimuddin and Z. N. Chen. Compact asymmetric-slit microstrip antennas for circular polarization. *IEEE Transactions on Antennas and Propagation*, 59(1):285–288, January 2011.

[16] B. Chen, Y. C. Jiao, F. C. Ren, and L. Zhang. Broadband monopole antenna with wideband circular polarization. *Progress In Electromagnetics Research Letters*, 32:19– 28, 2012.

[17] J. Wu, J. Ke, C. F. Jou, and C. Wang. Microstrip-fed broadband circularly polarised monopole antenna. *IET Microwaves, Antennas Propagation*, 4(4):518–525, April 2010.
[18] K. O. Gyasi, G. Wen, D. Inserra, Y. Huang, J. Li, A. E. Ampoma, and H. Zhang. A compact broadband cross-shaped circularly polarized planar monopole antenna with a ground plane extension. *IEEE Antennas and Wireless Propagation Letters*, 17(2):335–338, Feb 2018.
[19] A. Frotanpour and H. Hassani. A dual-broadband circularly polarized halved falcate-shape printed monopole antenna. *International Journal of RF and Microwave Computer-Aided Engineering*, 21(6):636–641, September 2011.
[20] C. Raviteja, C. Varadhan, M. Kanagasabai, A. K. Sarma, and S. Velan. A fractal-based circularly polarized UHF RFID reader antenna. *IEEE Antennas and Wireless Propagation Letters*, 13:499–502, 2014.
[21] S. Ahdi Rezaeieh, A. Abbosh, and M. A. Antoniades. Compact CPW-fed planar monopole antenna with wide circular polarization bandwidth. *IEEE Antennas and Wireless Propaga-tion Letters*, 12:1295–1298, 2013.
[22] J. Row. Design of aperture-coupled annular-ring microstrip antennas for circular polarization. *IEEE Transactions on Antennas and Propagation*, 53(5):1779–1784, May 2005.
[23] C. Su and J. Row. Slot-coupled microstrip antenna for broadband circular polarisation. Electronics Letters, 42(6):318–319, March 2006.
[24] J. Sze, C. G. Hsu, M. Ho, Y. Ou, and M. Wu. Design of circularly polarized annular-ring slot antennas fed by a double-bent microstripline. *IEEE Transactions on Antennas and Propagation*, 55(11):3134–3139, November 2007.
[25] J. Row. The design of a squarer-ring slot antenna for circular polarization. *IEEE Transactions on Antennas and Propagation*, 53(6):1967–1972, June 2005.
[26] S. H. Yeung, K. F. Man, and W. S. Chan. A bandwidth improved circular polarized slot antenna using a slot composed of multiple circular sectors. *IEEE Transactions on Antennas and Propagation*, 59(8):3065–3070, August 2011.
[27] R. Xu, J. Li, K. Wei, and G. Yang. A broadband slot antenna with unidirectional circularly polarized radiation patterns. *IEEE Antennas and Wireless Propagation Letters*, 16:317–320, 2017.
[28] H. Zhang, Y. Jiao, L. Lu, and C. Zhang. Broadband circularly polarized square-ring-loaded slot antenna with flat gains. *IEEE Antennas and Wireless Propagation Letters*, 16:29–32, 2017.
[29] T. Chang and J. Lin. Circularly polarized antenna having two linked slot-rings. *IEEE Transactions on Antennas and Propagation*, 59(8):3057–3060, August 2011.
[30] X. Q., Nasimuddin and Z. Ning Chen. Dual-square-ring-shaped slot antenna for wideband circularly polarized radiation. *Microwave and Optical Technology Letters*, 56(11):2645–2649, August 2014.
[31] T. Chang. Wideband circularly polarised antenna using two linked annular slots. *Electronics Letters*, 47(13):737–739, June 2011.

[32] S. Mohammadi, J. Nourinia, C. Ghobadi, J. Pourahmadazar, and M. Shokri. Compact broadband circularly polarized slot antenna using two linked elliptical slots for C-band applications. *IEEE Antennas and Wireless Propagation Letters*, 12:1094–1097, 2013.

[33] K. L. W. J. Yi Sze and C. C. Huang. Coplanar waveguide-fed square slot antenna for broadband circularly polarized radiation. *IEEE Transactions on Antennas and Propagation*, 51(8):2141–2144, August 2003.

[34] M. Shokri, V. Rafii, S. Karamzadeh, Z. Amiri, and B. Virdee. Miniaturised ultra-wideband circularly polarised antenna with modified ground plane. *Electronics Letters*, 50(24):1786–1788, 2014.

[35] P. Sadeghi, J. Nourinia, and C. Ghobadi. Square slot antenna with two spiral slots loaded for broadband circular polarisation. *Electronics Letters*, 52(10):787–788, 2016.

[36] Y. Ojaroudi, N. Ojaroudi, and N. Ghadimi. Circularly polarized microstrip slot antenna with a pair of spur-shaped slits for WLAN applications. *Microwave and Optical Technology Letters*, 57(3):756–759, January 2015.

[37] C. C. Chou, K. H. Lin, and H. L. Su. Broadband circularly polarised cross-patch- loaded square slot antenna. *Electronics Letters*, 43(9):485–486, April 2007.

[38] W. Liu, C. Wu, and Y. Tseng. Parasitically loaded CPW-fed monopole antenna for broadband operation. *IEEE Transactions on Antennas and Propagation*, 59(6):2415–2419, June 2011.

[39] X. Q. Zhang, Y. C. Jiao, and W. H. Wang. Compact dual-band dual-sense circularly-polarized CPW-fed slot antenna. *Progress In Electromagnetics Research Letters*, 34:197–205, 2012.

[40] T. Wu, J. Chen, and P.F. Wu. Compact triple-band CPW-fed square slot antenna with dual-polarization characteristics for wireless applications. *Progress In Electromagnetics Research C*, 105:241–251, 2020.

[41] J. Shen, C. Lu, W. Cao, J. Yang, and M. Li. A novel bidirectional antenna with broadband circularly polarized radiation in X-band. *IEEE Antennas and Wireless Propagation Letters*, 13:7–10, 2014.

[42] M. Nosrati and N. Tavassolian. Miniaturized circularly polarized square slot antenna with enhanced axial-ratio bandwidth using an antipodal Y-strip. *IEEE Antennas and Wireless Propagation Letters*, 16:817–820, 2017.

[43] R. Ma, and Q. Feng. Design of broadband circularly polarized square slot antenna for UHF RFID applications. *Progress In Electromagnetics Research C*, 111:97–108, 2021.

[44] R. Xu, J. Li, and J. Liu. A design of broadband circularly polarized C-shaped slot antenna with sword-shaped radiator and its array for l/s-band applications. *IEEE Access*, 6:5891–5896, 2018.

[45] Y. W. Liu and P. Hsu. Broadband circularly polarised square slot antenna fed by coplanar waveguide. *Electronics Letters*, 49(16):976–977, August 2013.

[46] J. Sze and S. Pan. Design of CPW-fed circularly polarized slot antenna with a miniature configuration. *IEEE Antennas and Wireless Propagation Letters*, 10:1465–1468, 2011.

[47] K. Saraswat and A. R. Harish. CPW fed wideband circularly polarized monopole antenna. In *2016 Asia-Pacific Microwave Conference (APMC)*, pp. 1–3, 2016.
[48] K. Saraswat, A. K. Awasthi, and A. R. Harish. A fishing hook shaped dipole antenna for broadband circular polarization. In 2018 3rd International Conference on Microwave and Photonics (ICMAP), pp. 1–2, 2018.
[49] K. Saraswat and A. R. Harish. Discontinuous annular ring monopole antenna for circular polarization. In 2019 IEEE Indian Conference on Antennas and Propogation (InCAP), pp. 1–3, 2019.
[50] C. Y. D. Sim, H. D. Chen, L. Zuo, and T. A. Chen. CPWfed sqaure ring slot antenna with circular polarization radiation for WiMAX/WLAN applications. Microwave and Optical Technology Letters, 57(4):886–891, February 2015.
[51] G. Augustin and T. A. Denidni. Coplanar waveguide-fed uniplanar trapezoidal antenna with linear and circular polarization. *IEEE Transactions on Antennas and Propagation*, 60(5):2522–2526, May 2012.
[52] J. Liu, Y. Li, Z. Liang, and Y. Long. A planar quasi-magnetic–electric circularly polarized antenna. *IEEE Transactions on Antennas and Propagation*, 64(6):2108–2114, June 2016.
[53] Y. Wen, B. Wang, and X. Ding. Wide-beam circularly polarized microstrip magneticelectric dipole antenna for wide-angle scanning phased array. *IEEE Antennas and Wireless Propagation Letters*, 16:428–431, 2017.
[54] S. X. Ta, I. Park, and R. W. Ziolkowski. Broadband electrically small circularly polarized directive antenna. *IEEE Access*, 5:14657–14663, 2017.
[55] K. Ding, C. Gao, T. Yu, and D. Qu. Broadband C-shaped circularly polarized monopole antenna. *IEEE Transactions on Antennas and Propagation*, 63(2):785–790, February 2015.
[56] K. Ding, C. Gao, Y. Wu, D. Qu, and B. Zhang. A broadband circularly polarized printed monopole antenna with parasitic strips. *IEEE Antennas and Wireless Propagation Letters*, 16:2509–2512, 2017.
[57] S. Pan, J. Sze, and P. Tu. Circularly polarized square slot antenna with a largely enhanced axial-ratio bandwidth. *IEEE Antennas and Wireless Propagation Letters*, 11:969–972, 2012.
[58] R. K. Saini and S. Dwari. CPW-fed broadband circularly polarized rectangular slot antenna with L-shaped feed line and parasitic elements. *Microwave and Optical Technology Letters*, 57(8):1788–1794, May 2015.
[59] J. Y. Sze and S. P. Pan. Design of broadband circularly polarized square slot antenna with a compact size. *Progress in Electromagnetics Research*, 120:513–533, 2011.
[60] U. Ullah and S. Koziel. A broadband circularly polarized wide-slot antenna with a miniaturized footprint. *IEEE Antennas and Wireless Propagation Letters*, 17(12):2454–2458, December 2018.
[61] J. Row and S. Wu. Circularly-polarized wide slot antenna loaded with a parasitic patch. *IEEE Transactions on Antennas and Propagation*, 56(9):2826–2832, September 2008.
[62] W. Yang and J. Zhou. Wideband circularly polarized cavity-backed aperture antenna with a parasitic square patch. *IEEE Antennas and Wireless Propagation Letters*, 13:197–200, 2014.

[63] Y. Zhang, Y. Jiao, Z. Zhang, and B. Li. Miniaturised CP aperture antenna with tri-mode operation for broadening bandwidth. *Electronics Letters*, 54(3):122–124, 2018.

[64] Y. Zhang, Y. Jiao, Z. Zhang, and S. Feng. Wideband accurate-out-of-phase-fed circularly polarized array based on penta-mode aperture antenna element with irregular cavity. *IEEE Transactions on Antennas and Propagation*, 67(1):638–642, January 2019.

[65] N. Kushwaha and R. Kumar. Compact coplanar waveguide-fed wideband circular polarised antenna for navigation and wireless applications. *IET Microwaves, Antennas Propagation*, 9(14):1533–1539, 2015.

[66] J. Sze and W. Chen. Axial-ratio-bandwidth enhancement of a microstrip-line-fed circularly polarized annular-ring slot antenna. *IEEE Transactions on Antennas and Propagation*, 59(7):2450–2456, July 2011.

[67] X. Jiang, Z. Zhang, Y. Li, and Z. Feng. Wideband unidirectional circularly polarised slot array with integrated feeding network. *Electronics Letters*, 50(15):1039–1040, July 2014.

[68] P. Xu, Z. H. Yan, T. L. Zhang, and X. Q. Yang. Broadband circularly polarized slot antenna array with fan-shaped feed line and L-shaped grounded strips. Progress In Electromagnetics Research Letters, 44:125–131, 2014.

[69] P. S. Hall. Application of sequential feeding to wide bandwidth, circularly polarised microstrip patch arrays. IEE Proceedings H – Microwaves, Antennas and Propagation, 136(5):390–398, October 1989.

[70] W. Yang, J. Zhou, Z. Yu, and L. Li. Bandwidth- and gain-enhanced circularly polarized antenna array using sequential phase feed. *IEEE Antennas and Wireless Propagation Letters*, 13:1215–1218, 2014.

[71] S. Mohammadi-Asl, J. Nourinia, C. Ghobadi, and M. Majidzadeh. Wideband compact circularly polarized sequentially rotated array antenna with sequential-phase feed network. *IEEE Antennas and Wireless Propagation Letters*, 16:3176–3179, 2017.

[72] X. Jiang, Z. Zhang, Y. Li, and Z. Feng. A low-cost wideband circularly polarized slot array with integrated feeding network and reduced height. *IEEE Antennas and Wireless Propagation Letters*, 15:222–225, 2016.

[73] J. Sze, C. G. Hsu, Z. Chen, and C. Chang. Broadband CPW-fed circularly polarized square slot antenna with lightening-shaped feedline and inverted-L grounded strips. *IEEE Transactions on Antennas and Propagation*, 58(3):973–977, March 2010.

[74] K. Saraswat and A. R. Harish. Analysis of wideband circularly polarized ring slot antenna using characteristics mode for bandwidth enhancement. *International Journal of RF and Microwave Computer-Aided Engineering*, 28(2):e21186, February 2018.

[75] K. Saraswat, T. Kumar, and A. R. Harish. A corrugated g-shaped grounded ring slot antenna for wideband circular polarization. *International Journal of Microwave and Wireless Technologies*, 12(5):431–436, June 2020.

[76] K. Saraswat and A. R. Harish. A coplanar waveguide fed tilted fractal inspired slot antenna for wideband circular polarization. *Microwave and Optical Technology Letters*, 61(6):1607–1611, June 2019.

[77] C. Y. Kang, S. L., H. Zong, Z. H. Zhao, and X. Y. Zhang. A wide-band circularly polarized wide-gap antenna loaded with a Y-shaped metal strip for L-band application. *International Journal of Antennas and Propagation*, 2015:1–12, 2015.

[78] J. Y. Sze and S. P. Pan. Design of broadband circularly polarized square slot antenna with a compact size. *Progress In Electromagnetics Research*, 120:513–533, 2011.

[79] C. Wang and C. Chen. CPW-fed stair-shaped slot antennas with circular polarization. *IEEE Transactions on Antennas and Propagation*, 57(8):2483–2486, Aug 2009.

[80] J. Sze and C. Chang. Circularly polarized square slot antenna with a pair of inverted-L grounded strips. *IEEE Antennas and Wireless Propagation Letters*, 7:149–151, 2008.

[81] R. V. S. Ramkrishna and Raj Kumar. A CPW-fed wideband circularly polarized slot antenna with FSS for enhanced gain. Microwave and Optical Technology Letters, 57(5):1199–1204, March 2015.

[82] K. Luo, B. Chen, Y. Yi, and Y. Wang. Miniaturised design of coplanar waveguide-fed broadband circularly polarised slot antenna. *IET Microwaves, Antennas Propagation*, 9(11):1200–1204, 2015.

[83] J. Yuan and Y. Li. A compact circularly polarized microstrip ring antenna using a slotted ground for GNSS applications. Progress In Electromagnetics Research Letters, 88:29–36, 2020.

[84] Y. Li, B. Tian, J. Xue, and G. Ge. Compact dual-band circularly polarized antenna design for navigation terminals. *IEEE Antennas and Wireless Propagation Letters*, 15:802–805, 2016.

[85] Z. Wang, R. She, J. Han, S. Fang, and Y. Liu. Dual-band dual-sense circularly polarized stacked patch antenna with a small frequency ratio for UHF RFID reader applications. *IEEE Access*, 5:15260–15270, 2017.

[86] Z. Liang, D. Yang, X. Wei, and E. Li. Dual-band dual circularly polarized microstrip antenna with two eccentric rings and an arc-shaped conducting strip. *IEEE Antennas and Wireless Propagation Letters*, 15:834–837, 2016.

[87] J. Li, J. Shi, L. Li, T. A. Khan, J. Chen, Y. Li, and A. Zhang. Dual-band annular slot antenna loaded by reactive components for dual-sense circular polarization with flexible frequency ratio. *IEEE Access*, 6:64063–64070, 2018.

[88] W. Li, B. Liu, and H. Zhao. The U-shaped structure in dual-band circularly polarized slot antenna design. *IEEE Antennas and Wireless Propagation Letters*, 13:447–450, 2014.

[89] W. Weng, J. Sze, and C. Chen. A dual-broadband circularly polarized slot antenna for WLAN applications. *IEEE Transactions on Antennas and Propagation*, 62(5):2837–2841, May 2014.

[90] C. Deng, Y. Li, Z. Zhang, G. Pan, and Z. Feng. Dual-band circularly polarized rotated patch antenna with a parasitic circular patch loading. *IEEE Antennas and Wireless Propagation Letters*, 12:492–495, 2013.

[91] Y. Shao and Z. Chen. A design of dual-frequency dual-sense circularly-polarized slot antenna. *IEEE Transactions on Antennas and Propagation*, 60(11):4992–4997, November 2012.

[92] M. Midya, S. Bhattacharjee, and M. Mitra. CPW-fed dual-band dual-sense circularly polarized antenna for WiMAX application. *Progress In Electromagnetics Research Letters*, 81:113–120, 2019.
[93] X. Rui, J. Li, and K. Wei. Dual-band dual-sense circularly polarised square slot antenna with simple structure. *Electronics Letters*, 52(8):578–580, 2016.
[94] K. Kandasamy, B. Majumder, J. Mukherjee, and K. P. Ray. Dual-band circularly polarized split ring resonators loaded square slot antenna. *IEEE Transactions on Antennas and Propagation*, 64(8):3640–3645, August 2016.
[95] R. Xu, J. Li, Y. Qi, Y. Guangwei, and J. Yang. A design of triple-wideband triple-sense circularly polarized square slot antenna. *IEEE Antennas and Wireless Propagation Letters*, 16:1763–1766, 2017.
[96] R. Xu, J. Li, J. Liu, S. Zhou, K. Wei, and Z. Xing. A simple design of compact dual wideband square slot antenna with dual-sense circularly polarized radiation for WLAN/Wi-Fi communications. *IEEE Transactions on Antennas and Propagation*, 66(9):4884–4889, September 2018.
[97] Y. Chen, Y. Jiao, G. Zhao, F. Zhang, Z. Liao, and Y. Tian. Dual-band dual-sense circularly polarized slot antenna with a C-shaped grounded strip. *IEEE Antennas and Wireless Propagation Letters*, 10:915–918, 2011.
[98] K. Saurav, D. Sarkar, A. Singh, and K. V. Srivastava. Multiband circularly polarized cavity-backed crossed dipole antenna. *IEEE Transactions on Antennas and Propagation*, 63(10):4286–4296, October 2015.
[99] J. D. Zhang, L. Zhu, N. W. Liu, and W. Wu. Dual-band and dual-circularly polarized single-layer microstrip array based on multi resonant modes. *IEEE Transactions on Antennas and Propagation*, 65(3):1428–1433, March 2017.
[100] S. X. Ta, K. Lee, I. Park, and R. W. Ziolkowski. Compact crossed-dipole antennas loaded with near-field resonant parasitic elements. *IEEE Transactions on Antennas and Propagation*, 65(2):482–488, February 2017.
[101] Z. N. C. Nasimuddin and X. Qing. Dual-band circularly polarized S-shaped slotted patch antenna with a small frequency-ratio. *IEEE Transactions on Antennas and Propagation*, 58(6):2112–2115, June 2010.
[102] X. L. Bao and M. J. Ammann. Monofilar spiral slot antenna for dual-frequency dual-sense circular polarization. *IEEE Transactions on Antennas and Propagation*, 59(8):3061–3065, August 2011.
[103] Y. K. Jung and B. Lee. Dual-band circularly polarized microstrip RFID reader antenna using metamaterial branch-line coupler. *IEEE Transactions on Antennas and Propagation*, 60(2):786–791, February 2012.
[104] W. T. Hsieh, T. H. Chang, and J. F. Kiang. Dual-band circularly polarized cavity-backed annular slot antenna for GPS receiver. *IEEE Transactions on Antennas and Propagation*, 60(4):2076–2080, April 2013.
[105] S. Kibria, M. T. Islam, and B. Yatim. New compact dual-band circularly polarized universal RFID reader antenna using ramped convergence particle swarm optimization. *IEEE Transactions on Antennas and Propagation*, 62(5):2795–2801, May 2014.

[106] C. Chen and E. K. N. Yung. Dual-band dual-sense circularly-polarized CPW-fed slot antenna with two spiral slots loaded. *IEEE Transactions on Antennas and Propagation*, 57(6):1829–1833, June 2009.
[107] C. H. Chen and E. K. N. Yung. Dual-band circularly-polarized CPW-fed slot antenna with a small frequency ratio and wide bandwidths. *IEEE Transactions on Antennas and Propagation*, 59(4):1379–1384, April 2011.
[108] K. Saraswat and A. R. Harish. Dual-band CP coplanar waveguide-fed split-ring resonator-loaded g-shaped slot antenna with wide-frequency ratio. *IET Microwaves, Antennas & Propagation*, 12(12):1920–1925, October 2018.
[109] K. Saraswat and A. R. Harish. A dual band circularly polarized 45° rotated rectangular slot antenna with parasitic patch. *International Journal of Electronics and Communications*, 123(153260):153260, August 2020.
[110] K. Saurav, D. Sarkar, and K. V. Srivastava. Dual-band circularly polarized cavity-backed crossed-dipole antennas. *IEEE Antennas and Wireless Propagation Letters*, 14:52–55, 2015.
[111] T. M. Parvathy and V. G. Ajay. Circularly polarized split ring resonator loaded slot antenna for RFID readers and WLAN applications. *Advanced Electromagnetics*, 7(5):1–6, September 2018.

Chapter 4

Multi-input–multi-output antenna

Reconfigurable antenna

Akanksha Singh, Lipsa Das, and Pooja Anand

4.1 INTRODUCTION

Ultrawideband (UWB) approaches have attracted considerable attention since the Federal Communications Commission (FCC) made the spectrum from 3.1 to 10.6 GHz available for commercial purposes. This is a result of features like high data rates, wide bandwidth, and affordable prices. One of the essential elements of a UWB wireless communication system, UWB antennas, has been the subject of extensive research. Despite the large frequency bands that the UWB communication system uses, the allowable power spectral density of the UWB signal is quite constrained (Kaiser et al., 2009). By allowing data to be transmitted over several channels, multiple-input and multiple-output (MIMO) systems boost channel capacity without requiring more power (Wallace et al., 2003). Additionally, MIMO antennas can be employed to achieve diversity performance and reduce multipath fading's effects (Foschini et al., 1998). Modern wireless communication standards like Long-Term Evolution (LTE), wireless local area network (WLAN; IEEE 802.11n), and Worldwide Interoperability for Microwave Access (WiMAX) rely heavily on the MIMO technology (WiMAX). MIMO technology can significantly increase network speed, capacity, and coverage without requiring more bandwidth, according to studies (Bolin et al., 2005; Foschini & Gans, 1998).

Portable wireless terminals have also been designed to handle various standards on a smaller device and offer a variety of features. Therefore, it is essential to provide small, multiband antennas to support wireless terminals with MIMO functionality. It was recently reported by Sorin et al. (2011) that various antenna monopole antennas are designed for the 2.5–2.7 GHz range and employed neutralization lines for decoupling. A two-band printed antenna was studied by Ding et al. (2007) using WLAN operations. The antenna is made up of two side-by-side monopoles that are printed on a printed circuit board (PCB) and have a number of ground planes to improve isolation and change the resonance frequency. In (Cui et al., 2011) a two-element folded monopole antenna for operation at 2.4 and 5.6 GHz was

DOI: 10.1201/9781003470854-4

suggested. To improve the isolation, two transmission lines are printed on the FR4 substrate and two slots are etched on the bottom surface. For mobile terminals, multiband antennas have also been created, such as Bhatti et al. (2009) for the 2.4/5.25/5.8-GHz bands and Li et al. (2011) for the 2.4/5.2/5.8-GHz WLAN, 2.5/3.5/5.5-GHz WiMAX, and on the lower side 3.1–4.8 GHz UWB band. To decrease coupling and enhance impedance matching, two curved slots were carved onto the ground plane (Li et al., 2011). Lower-frequency MIMO antenna systems for GSM 850/900/DCS/UMTS/LTE2500 and GSM 900/1800/1900/UMTS, respectively, have recently been proposed in Yang et al. (2012) and Zeng et al. (2012). For usage in MIMO terminals, various designs have been put out in the literature.

Multi-element antennas have the potential to significantly boost channel capacity and boost the dependability of UWB wireless communication applications. However, the UWB MIMO system's performance will suffer from significant mutual coupling brought on by the antenna elements. As a result, one of the major difficulties in MIMO antenna designs is isolation augmentation. Numerous decoupling methods have been documented up to this point (Mak et al., 2008). In (Li et al., 2013) the mutual coupling can be decreased by utilizing various decoupling elements, such as floating parasitic (Zhang et al., 2009) and tree-like (Khan et al., 2014) architectures. However, the constructions in Khan et al. (2014) and Gallo et al. (2012) take up a lot of room and complicate the antenna layout. Due to their orthogonal emission patterns, diverse antenna also represents a promising option for producing low mutual interaction between antenna parts (Gallo et al., 2012; Ren et al., 2014). Additionally, antennas are needed to filter out the undesirable band to reject interference with the current wireless communication systems, like the WLAN running at 5.15–5.85 GHz. Many UWB MIMO antennas that operate in the band-reject mode have recently been introduced (Qin et al., 2022; Zhao et al., 2012a). The split-ring resonator (SRR) slot (Ren et al., 2014), the open stub (Qin et al., 2022), and the arc-shaped slot are introduced into the antenna element, respectively, to achieve the band-notched features. Although good performance may be achieved, the proposed antenna's smallest dimension is greater than the antennas with an area of 42.5 mm × 42.5 mm × 18 mm (Qin et al., 2022).

The channel capacity and dependability of wireless communication systems have also been enhanced using MIMO technology (Foschini & Gans, 1998). A variety of antennas for MIMO systems have recently been introduced (Chung & Yoon, 2007). However, due to both strongly coupled antennas and coupling currents of the surface on the ground plane, particularly in a small mobile device, maintaining good isolation between the internal antennas of the MIMO system is challenging. Additionally, the mutual coupling decreases the radiation efficiency and raises the envelope correlation of the two ports (Chung & Yoon, 2007; Diallo et al., 2006). Putting antenna array a half-wavelength apart will result in the lowest coupling between two antennas. Due to the commercial products' constrained area, it is challenging to implement multiple antennas for MIMO systems.

Another goal is to improve the isolation between the components of the antenna. The antenna elements should typically have directional gains. It should be noted that the approaches used to eliminate mutual coupling should have little effect on wideband impedance matching for UWB applications. Numerous strategies have been put forth to address the aforementioned difficulties. Three main categories can be used to group these decoupling strategies. Utilizing UWB diversity antennas is the first technique (Mohammad et al., 2013; Yoon et al., 2011; Zhang et al., 2012a). This technique works on a similar principle as dual-polarized antennas. Low coupling between the antenna elements is possible because of the orthogonality of the gain characteristics of the antenna elements. Additionally, the MIMO antennas for UWB applications employing this technique are often small in size. The second method makes use of decoupling structures (Hong et al., 2008; Lee et al., 2012; Liu et al., 2006; Zhang et al., 2009), including parasitic meander lines (Lee et al., 2012) and treelike structures (Zhang et al., 2009). The antennas using this technique might achieve a high level of isolation. However, because decoupling structures typically have complex actual forms, it is challenging to achieve a reduction of the antenna's dimensions. The third way is the most popular and may be thought of as a mix of the first two methods (Gallo et al., 2012; Gao et al., 2014; Li et al., 2013; See & Chen, 2009). Compact size and minimal mutual coupling are typically achievable by using UWB diversity antennas and decoupled components between both the antenna elements at the same time.

A reconfigurable antenna is a type of antenna that can dynamically change its operating characteristics, such as frequency, radiation pattern, polarization, or beam direction. It offers the flexibility to adapt to different communication requirements and environmental conditions. The reconfigurability of these antennas is typically achieved using various techniques, such as mechanical, electrical, or magnetic reconfiguration. These techniques allow the antenna to modify its physical structure or electrical properties in real time or in response to control signals.

4.2 TYPES OF ANTENNA SYSTEM

4.2.1 Microstrip antenna

Metallic strips or patches put on an FR4 substrate are microstrip antennas as shown in Figure 4.1. Due to their two-dimensional nature, microstrip antennas are easy to design and produce. With the aid of contemporary printed circuit technology, they are simple to make. Low-profile microstrip antennas may adhere to both planar and non-planar surfaces. These antennas support both circular and linear polarization. These antennas come in a variety of shapes,

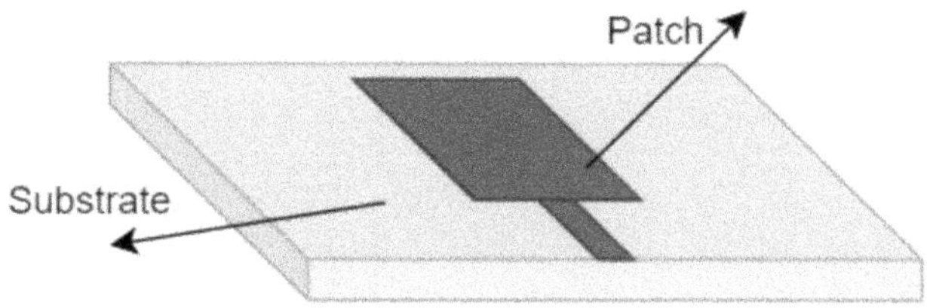

Figure 4.1 Microstrip antenna structure.

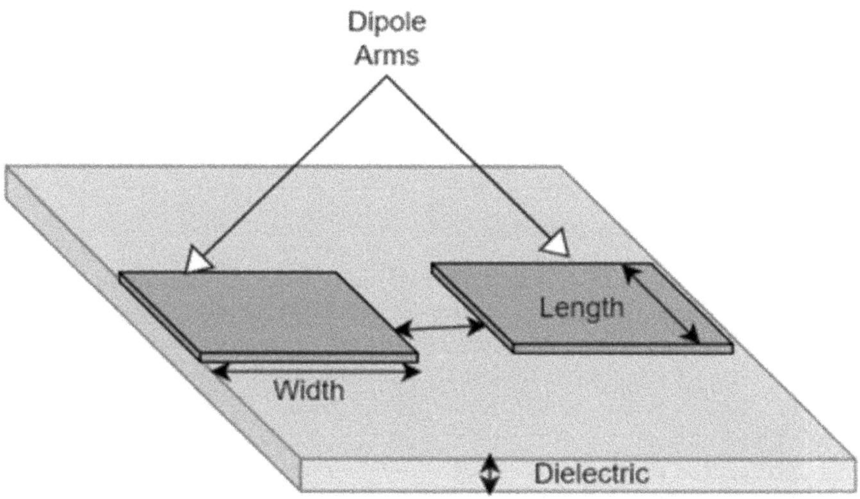

Figure 4.2 Printed dipole antenna.

including rectangle, square, circular, triangle, and elliptical patterns, and can be simply installed on hard surfaces. The majority of Global Positioning System (GPS) devices employ a microstrip or patch antenna.

4.2.2 Printed dipole antennas

The printed dipole antenna, as shown in Figure 4.2, is simple in construction, having low cost, polarization diversity, large frequency coverage, and they are widely used in wireless communication applications. In the printed dipole antenna, two arms are constructed on both sides of the substrate. Dipole antennas are difficult to use in situations with limited space because of their size. Furthermore, issues with compact form-factor designs could arise from the deterioration of omnidirectional radiation patterns and the probable requirement for a balun. Printed dipole antennas are frequently employed in mm-wave and wireless communication applications.

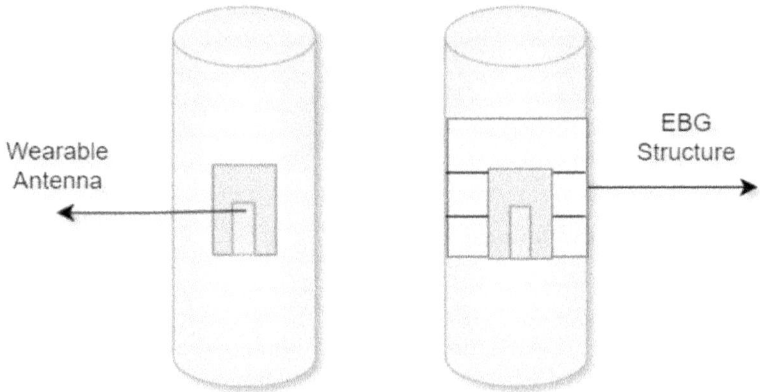

Figure 4.3 Monopole antenna.

4.2.3 Monopole antennas

Monopole antennas are very often positioned over a ground plane and are half as big as dipole antennas. Monopole antennas are shown in Figure 4.3 and are perfect applications where low-profile antenna is required due to its small size. If monopole antennas are positioned over high-impedance surfaces, they perform well in terms of radiation high-impedance surface (HIS). The fundamental criteria for wearable antennas are met by monopole antennas as they are low profile, inexpensive, and simple to construct. Monopole antennas are perfect for incorporation into clothing due to their straightforward, lightweight design.

4.2.4 Printed loop antennas

A printed loop antenna is constructed from one or more loops that can be in the form of a circle, square, or any other geometrical shape. The loop antenna as shown in Figure 4.4 ensures that the current in the loop stays in phase by having a diameter smaller than a wavelength. These antennas are lightweight and have a straightforward, small design. There are two different types of loop antennas: large loop antennas and small loop antennas. The small loop antennas are primarily utilized for receiving, whereas the large loop antennas are used for both transmission and reception. These antennas are perfect for compact radio devices and military-grade wearable communication systems. However, there are some drawbacks of the loop antenna as it has poor efficiency because excessive current flow causes power loss in the form of heat.

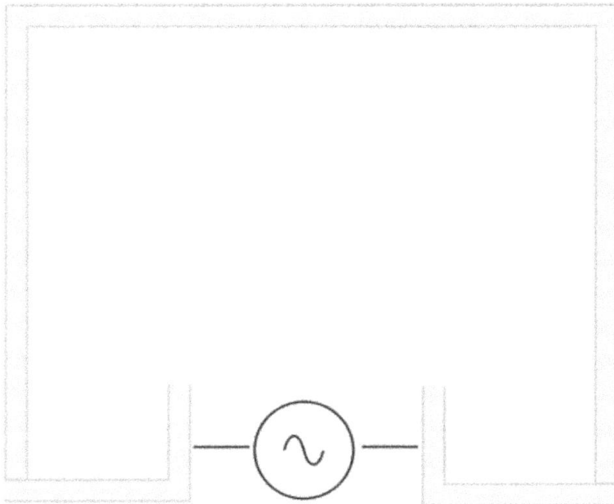

Figure 4.4 Printed loop antenna.

4.2.5 Slot antennas

An antenna having narrow, fine slots, as shown in Figure 4.5, is called a slot antenna. Typically utilized between 3 GHz and 24 GHz, slot antennas are exceedingly adaptable. This antenna features linear polarization and omni-directional radiating patterns. The slot's dimensions, shape, and material properties define the antenna's performance characteristics. Small form-factor wearables applications can benefit from its basic construction and flexibility. The antenna is perfect for medical and military purposes because it is simple to use on flexible surfaces like denim. Even when a person's position changes, this antenna nevertheless delivers reliable wireless data transmission.

4.2.6 Planar inverted-F antennas (PIFA)

Most of the applications for the Planar Inverted-F antenna (PIFA) are in portable smart gadgets. As implied by the name, these antennas have the shape of an inverted 'F' as shown in Figure 4.6. The antenna is favored by designers of wearable technology because of its reduced size and omni-directional pattern. PIFAs can also be printed using the same method as microstrips, which enables the printing of antennas on substrates and circuit boards. The PIFAs are ideal for body worn electronic devices due to its small size, multiband capabilities, and excellent specific absorption rate

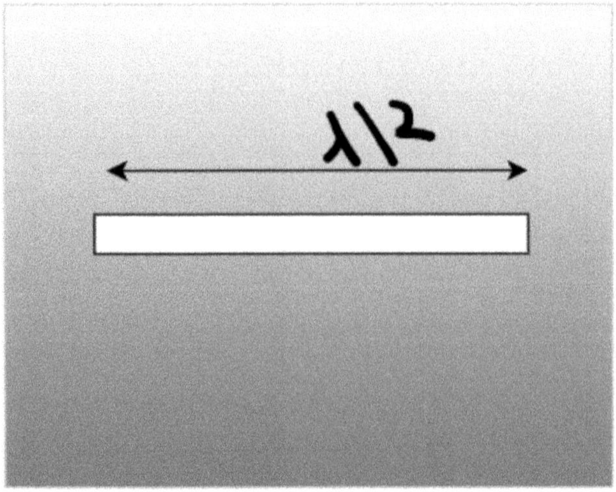

Figure 4.5 Slot antenna.

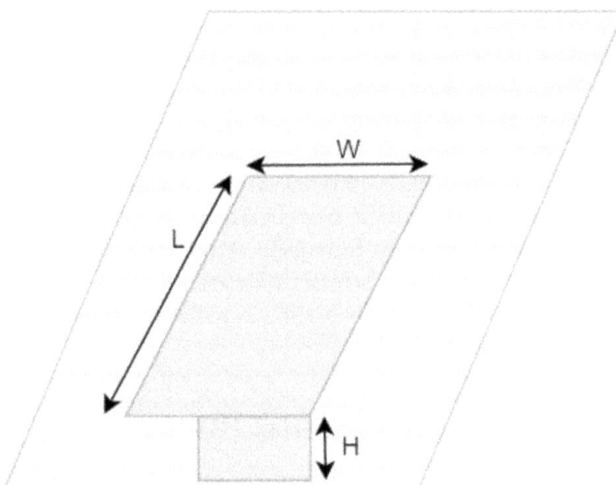

Figure 4.6 Planer inverted-F antenna.

(SAR) values. SAR measures energy absorption per unit mass by the human body when exposed to a radio frequency electromagnetic field, measured in W/kg. The FCC of the United States has established a 1.6 watts per kilogram (1.6 W/kg) SAR exposure limit for the public.

4.3 MIMO

MIMO can distribute the similar data as many signals simultaneously across multiple antennas while still using a single channel. An RF link's signal quality and strength are increased using various antennas in this instance of antenna diversity. At the transmission point, the data is divided into various data streams that are then recombined by a second MIMO radio set-up with the same of antennas on the receiving side. The receiver is built to account for any interference between the signals, lost signals, and even the little lag in timing between each signal's receipt. Traditional single antenna setups (SISO: Single In, Single Out) cannot deliver the same data on several streams. Due to this disadvantage of SISO, MIMO systems can send the same data on the single channel, and due to this it provides a number of advantages over conventional SISO designs like higher data rate and high channel capacity.

4.4 BASICS OF MIMO SYSTEMS

Fading may affect a channel, which will influence the ratio of signal to noise. In turn, if digitized data is being transferred, this will have an effect on the error rate. The diversity principle calls for sending the receiver various iterations of the same signal. The likelihood that they will all be impacted simultaneously is significantly decreased if these can be arranged to be impacted by the signal path in distinct ways. Diversity thus enhances performance and lowers mistake rates while stabilizing a link.

There are numerous and various modes that are available and offer the following benefits:

- Time diversity: By employing distinct timeslots and channel coding, an information can be broadcast at many times using time diversity.
- Various frequencies are used in this type of diversity. They might include the process of using various channels or tools like spread spectrum or orthogonal frequency-division multiplexing (OFDM).
- The foundation of multi-input–multi-output system is space diversity, understood in its fullest sense. It employs antennas placed in various locations to benefit from the many radio pathways that are present in a typical terrestrial environment.

As MIMO utilizes antenna arrays at the receiver and transmitter as shown in Figure 4.7, it selects different signal routes for each antenna to enable the usage of multiple signal paths, and it is essentially a radio antenna technology.

Key idea of the MIMO system is space–time signal processing. Considering this, MIMO antenna system can be taken as a logical development of the

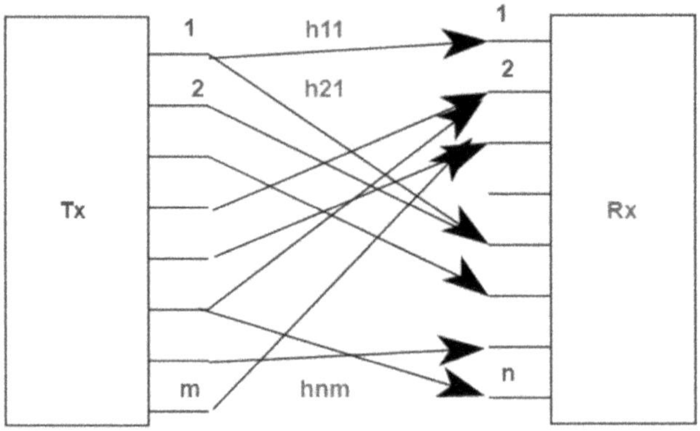

Figure 4.7 MIMO technology.

smart antenna systems that have been used for many years to enhance wireless system performance. Here, between the transmitter and the receiver, the signal might travel through number of different pathways. Moreover, the paths used by the signals will alter if the distance between the antenna elements is changed.

These multiple paths served to cause interference of the signal between the transmitter and receiver antenna systems. These extra routes can be effectively utilized by MIMO technology. By enhancing the ratio of signal to noise or expanding the link's data capacity, they can be employed to provide the radio link more robustness.

Multiple antennas are used in MIMO wireless technology, which allows for a significant increase in channel capacity while maintaining Shannon's law. With every pair of additional antennas added to the system, the channel's throughput is boosted linearly. MIMO wireless technology has thus emerged as one of the most significant wireless technologies employed in recent years. As spectral frequency band for radio communications grows into a more valuable resource, techniques are needed to make better use of the available bandwidth.

As MIMO system receives and combines numerous streams of the same data that are received at slightly varying time intervals, MIMO radios can actually boost signal strength even without clear line-of-site using the bounced and reflected RF signals. This concept is known as multipath fading. It is helpful in urban settings where signal deterioration among single antennas without a clear line of sight is a significant problem. Using MIMO technology, we are benefitted in the following ways.

1. Overall transmission rate can be enhanced, which lets more high-quality video or other information be transmitted over the difference channels.

2. Using several data streams, disadvantages like fading, loss of data packets can be minimized, which results in better video or audio quality.

4.5 ANTENNA TERMINOLOGIES

The various antenna technology configurations can have one or more inputs and outputs. In this scenario, the transmit antenna serves as the message signal by transmitting into the link or signal path, while the receive antenna serves as the output. It is at the wireless link's output.

Consequently, the different types of single and multiple antenna systems are defined as follows:

- SISO stands for single input single output
- SIMO – single input multiple output
- MISO stands for multiple input and single output
- MIMO stands for multiple input multiple output.

4.5.1 SISO

SISO antenna systems are essential in wireless communication, thus providing simple yet efficient transmission and receiving for various applications like satellite links and smartphones. They are adaptable, affordable, and reliable for various wireless devices and networks, but lack diversity and spatial multiplexing capabilities of multiple antenna systems. SISO is the simplest type of radio link in MIMO terms. This transmitter and receiver both use only one antenna, as shown in Figure 4.8, so it is effectively a normal radio channel. No diversity exists, and no additional processing is needed.

4.5.2 SIMO

When on the transmitter side we have a single antenna and on the receiver side we have several antennas as shown in Figure 4.9; this type of technique is known as SIMO, or single input multiple output. Additionally called as receive diversity, this is often employed to counteract the fading effect in receiver systems that receive signals from numerous liberated sources.

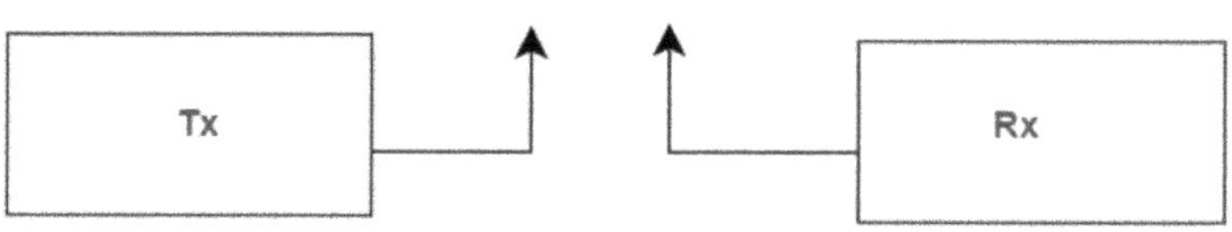

Figure 4.8 Single input single output antenna system.

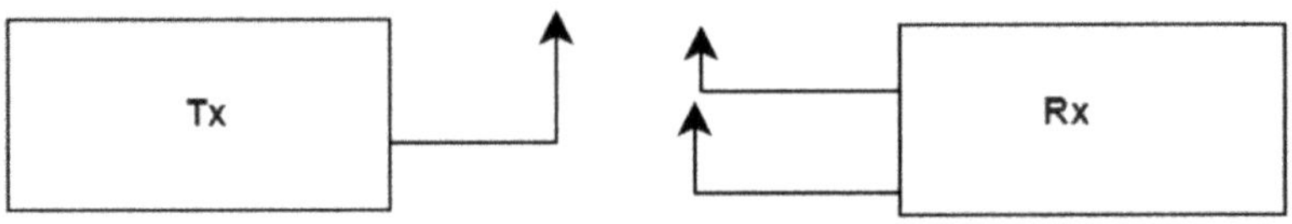

Figure 4.9 Single input multiple output antenna system.

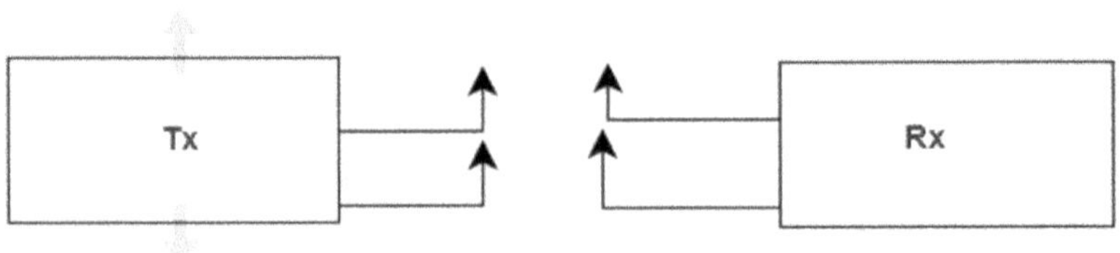

Figure 4.10 Multiple input multiple output antenna system.

SIMO antenna configuration enhances wireless communication systems by using multiple antennas for reception and a single antenna for transmission, improving signal quality, reliability, and data throughput. Used in cellular networks, Wi-Fi routers, and 5G technologies, it combats channel impairments and provides robust connectivity in diverse environments.

4.5.3 MIMO

MIMO refers to a radio link that has more than one antenna at each end as shown in Figure 4.10. Improvements in channel throughput and robustness can both be obtained with MIMO. MIMO antenna systems are crucial for modern wireless communication, enhancing spectral efficiency, data rates, and system performance. They use spatial diversity and multipath propagation to enable simultaneous communication with multiple users or data streams over the same frequency band. MIMO applications span cellular networks, Wi-Fi, and emerging technologies like 5G, delivering faster speeds, greater coverage, and robust connectivity.

4.6 IMPORTANCE OF MIMO

MIMO technology enhances wireless communication by achieving higher data rates by transmitting multiple data streams simultaneously over the same frequency band, resulting in faster downloads, faster uploads, and reduced latency. Wi-Fi networks, cellular fourth-generation (4G) LTE, and fifth-generation (5G) technologies all make use of MIMO technology, which is used widely in industries such as law enforcement, broadcast TV production, and government. In addition, it may be utilized in WLANs and is compatible with any wireless gadget that can use the 802.11n standard. MIMO is frequently employed in high-bandwidth communications when isolation from other microwave or RF systems is critical. First responders, for example, rely on it frequently because they cannot always rely on mobile networks in times of emergency, power outages, or network congestion.

MIMO technology enhances wireless communication by achieving higher data rates, improving coverage and reliability, and increasing capacity. It supports multiple users or data streams concurrently, allowing for better network congestion management. MIMO also improves performance in non-line-of-sight conditions, ensuring consistent performance in indoor environments. As wireless technologies evolve, MIMO antennas provide a foundation for implementing advanced features like beamforming, massive MIMO, and millimeter-wave communication, which promise higher data rates, lower latency, and better spectral efficiency, paving the way for transformative applications like augmented reality, autonomous vehicles, and Internet of Things (IoT) connectivity. MIMO technology enhances wireless communication by providing faster speeds, improved reliability, and enhanced network performance, benefiting both personal and professional settings.

4.7 RECONFIGURABLE MIMO ANTENNAS

Reconfigurable antennas are innovative wireless communication technologies that can dynamically adjust their parameters to suit changing environmental conditions or user needs. They optimize performance, enhance signal coverage, and mitigate interference in real time, making them ideal for mobile, satellite, radar, and IoT devices. This flexibility enhances spectrum utilization, connectivity, and user experiences.

An antenna that can dynamically, controllably, and irreversibly change its frequency and radiation characteristics is said to be reconfigurable. Reconfigurable antennas incorporate an inner mechanism (such as RF switches, varactor diodes, mechanical actuators, or tunable materials) to provide a dynamic response. This internal mechanism enables the purposeful rerouting of RF currents across the transmitter surface and produces reversible adjustments to the antenna's characteristics. Smart antennas have an external beamforming network, whereas reconfigurable antennas have a reconfiguration mechanism built in into the antenna itself. To maximize performance of the antenna in a changing environment or to meet shifting operational needs, reconfigurable antennas are utilized. Different reconfigurable techniques are shown in Figure 4.11.

Due to its property to provide dynamic frequency, pattern and polarization diversity on the one combined system with the benefit of enhanced

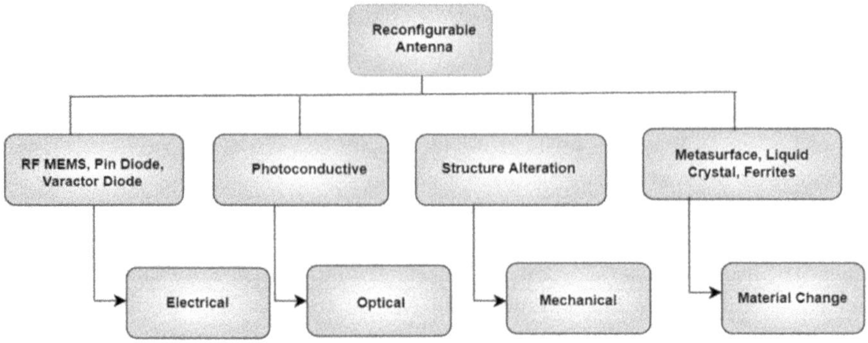

Figure 4.11 Antenna reconfigurable techniques.

signal reception, decreased space, and low cost, reconfigurable antennas have recently expanded the development of products. On the antenna plane, switching components including micro-electromechanical switch (MEMS), varactor, and PIN diodes have been used to create reconfigurable antennas. Polarization diversity antennas are also advantageous in satellite, mobile, and wireless communications systems because they help reduce fading brought on by multipath situations and bad weather. In MIMO systems, implementing polarization reconfigurability to boost channel capacity is also appealing. To achieve polarization, frequency, and pattern reconfiguration, reconfigurable microstrip patch antennas, slot antennas, and monopole antennas can be utilized.

Reconfigurable antennas are versatile devices that can be adjusted to suit specific applications. They can be frequency-adjustable, beam-reconfigurable, polarization-reconfigurable, pattern-reconfigurable, impedance-reconfigurable, or multiband-reconfigurable. Each type offers unique capabilities and benefits, allowing for flexibility, adaptability, and performance optimization. Frequency-adjustable antennas can adjust their operating frequency or resonate at multiple frequencies, while beam-reconfigurable antennas use beamforming techniques to focus energy towards desired areas. Polarization-reconfigurable antennas can switch between different polarization states, improving signal robustness and reliability. Pattern-reconfigurable antennas can alter radiation pattern characteristics, while impedance-reconfigurable antennas can adjust input impedance to match load conditions. Multiband-reconfigurable antennas can operate across multiple frequency bands.

A frequency reconfigurable antenna is greatly desired because it is practical for wireless communication systems that use multiple frequencies and different standards. However, by adjusting the radiation pattern, the antenna can significantly improve the performance of wireless systems by reducing source noise, enhancing security, and conserving energy by improving signal directions. The goal of frequency reconfiguration can be achieved via a few different control mechanisms. Electronic switches can be used to add or subtract length from the antenna, changing the effective length. For instance, PIN diodes were used to create tunable frequency bands on the fractal dipole antenna in Hamzah et al. (2012). Similar methods utilizing PIN diodes have also been documented in Lai et al. (2009), Majid et al. (2012a), Majid et al. (2012b). Reports on the application of RF MEMS for frequency tuning may be found in Erdil et al. (2007) through Pourziad et al. (2013). Utilizing varactor diodes or other devices with changeable reactive loading, which is often capacitance, is another way to alter the antenna's effective length. This enables seamless frequency tuning when the capacitance value is changed. In Abu Tarboush et al. (2011), an H-shaped antenna with varactor diode-based frequency selectivity was introduced. The same strategy has also been documented in Ren et al. (2013). Larger frequency shifts can also be delivered through structural or mechanical alterations. The physical design, which comprises the antenna's size and the actuation mechanism, presents the main difficulty in using these approaches. These antenna types have been described in Ruvio et al. (2007). Through structural, mechanical, electrical, and material changes, patterns can be reconfigured. Electrical changes are frequently utilized today as they require less maintenance than mechanical ones and are more compact. Electrical approaches use electronic switches such as PIN diodes, RF MEMS, and field-effect transistors (FETs). To induce radiation reconfiguration through electrical changes, there are three typical techniques. The parasitic tuning approach involves electrically switching or tuning parasitic elements. In Yang et al. (2007), a pattern-reconfigurable Yagi patch antenna was described. Similar methods have been explored in Zhang et al. (2005), where pattern reconfigurations were accomplished using parasitic tuning. Element phase shifting is a typical method for pattern reconfiguration. Beam steering is possible in this case because of the integration of phase shifts into array elements. In Gokalp and Civi (2011), a beam-steerable meander line antenna was introduced. Additionally, Morabito et al. (2012) and Ali et al. (2009) have explored a similar strategy. Utilizing a co-located radiating aperture is the third strategy for pattern reconfiguration. Pattern reconfiguration is accomplished using two switches to transition from slot line to coplanar waveguide (CPW) feeds, as described in Li et al. (2011). The same strategy was previously presented in Ren et al. (2014).

4.8 MIMO ANTENNA FOR IOT APPLICATIONS

Patch antennas are typically employed in the IoT and have GPS capabilities. A perturbation element in a patch antenna can be switched between linear, Left-Hand Circular Polarization (LHCP), or : Right-Hand Circular

Polarization (RHCP) polarizations by employing a PIN diode, varactor diode, or RF-MEMS switch. However, many patch antennas still only work in one of these polarizations. It can be challenging to select the ideal polarization for a transmission many times. The benefit of IoT devices with embedded antennas, such as chips and PCBs, is that they can accommodate more antennas in a smaller space, reducing the size of a sensor node. Comparable chip-based antennas do not achieve the gains of PCB antennas made of conductive traces. As the perception layer makes it easier to access and record all of the system's data, an effective system is basically needed to detect signal data with an effective data rate control. For PCB antennas, a variety of antenna topologies are possible, including inverted-F, L, and folded monopole (Mathur & Dwari, 2021). Additionally crucial to the advancement of chip antennas is the ground plane. A ground plane with a narrower bandwidth and a better radiation pattern can greatly restrict the design. Embedded chip antennas take up a lot of room on the circuit board and, like any antenna, their gain is directly related to the radiator's volume. However, sensor nodes with a PCB antenna have the advantage of maintaining a consistent shape, which may make mounting and enclosing them in any environment easier. Sectorial and Yagi antenna elements can be used for IoT base stations or to increase the radiation range. High-gain directional antennas have a longer communication range than omnidirectional antennas due to mechanical fluctuation. Additionally, effective communication and control in an IoT system require risk assessments based on design primitives. The physical layer sensor properties in the three categories of aristocratic, reliable, and security are sequentially elicited by the perception layer with a high chance of false triggering danger. The same significant threat exists for channels of communication and remote access in physical or virtual integrated development environments (IDE). The end user interface (EUI), which directs all RF stimuli and elicits challenging qualities to deal with RF access risk attributes, serves as the decision trigger. As a result, both physical and virtual layers of IoT systems will eventually require secure sensor properties and effective RF perception. In IoT devices, the system dependability can be improved and the risk of unwanted interference can be reduced by an antenna's suitable reconfigurability.

Both 5G and the IoT will demand extremely high data transfer rates. MIMO technology with beamforming is one of the most important transmission paradigms for supercharged 5G networks and the IoT. On top of the transmission tower, a number of antennae are going to be mounted. It will be able to zero in on a certain user at a specific area and communicate with that person while simultaneously making use of a number of antennae.

When the user points the antennas in a certain direction, the antennas will be able to keep track of the user and any changes in where the user is. It enables for network administrators to keep their service going without any problems. MIMO can help IoT systems, building automation, smart cities, and connected cars.

4.9 RECONFIGURABLE MIMO ANTENNA FOR IOT: A CASE STUDY

Reconfigurable antennas for CR applications have attracted increased interest in recent years. Designing tunable wideband to narrowband antennas or an UWB sensing antenna in addition to tenable narrowband antennas is the focus of the research. A p-i-n diode is used to switch between the wideband and narrowband states, and two varactor diodes mounted on a bandpass filter are used to control narrowband tuning. These switches are based on Gallium Arsenide (GaAs) field effect transistors. In p-i-n diodes, the varactor diodes, and the MEMS are often used in antennas to change the frequency. For IoT devices, the next technologies for wireless communication will need to have relatively high bandwidth utilization and efficiency in a fading environment.

In this design, four circular ring-shaped patch antennas are combined with the feed line for the tuning of frequency. For tuning purposes, a frequency configurable antenna is presented in Mathur & Dwari, 2021).

4.9.1 Antenna design and operation

Figure 4.12 depicts the configuration of the 4-port UWB-MIMO antenna. Its overall dimensions are $90 \times 90 \times 1.6$ mm^3, and its FR4 substrate is 1.6 mm thick (loss tangent is 0.02, dielectric constant $\delta = 4.4$). Four

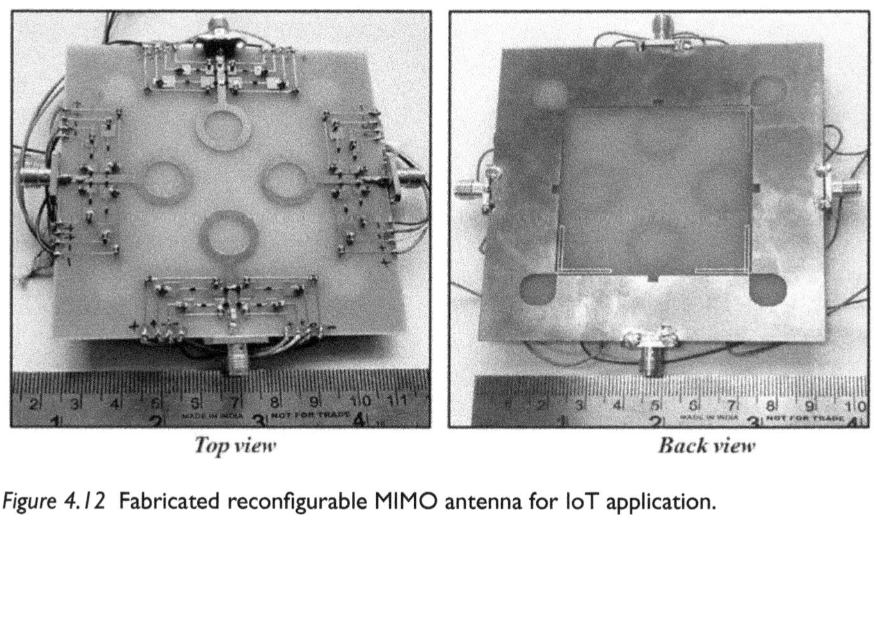

Top view *Back view*

Figure 4.12 Fabricated reconfigurable MIMO antenna for IoT application.

circular ring-shaped radiators are in the top layer and are driven by 50 Ω microstrip lines. To achieve polarization diversity and reduce coupling, adjacent radiators are positioned orthogonally to one another. Each antenna element's UWB response is optimized on a partial ground plane on the substrate's bottom side. At the four corners of the ground plane, four teardrop-shaped slots connected to foldable meandered strips are designed to reduce mutual interference between two adjacent ports. All other ports' answers are identical because of symmetry. The antenna has port isolation that is higher than 15 dB for the majority of the frequency range, which ranges around 3–12 GHz.

4.9.2 Reconfigurable antenna design

The adjustable stubs as discussed in the previous section are merged with the UWB-MIMO antenna in the final design. The fabricated antenna is shown in Figure 4.12 in its final structure, which consists of four ring-shaped monopole antenna elements, with two sets of SIRS with DC bias networks for p-i-n diodes, and varactor diodes for each monopole. PIN diodes positioned in between the microstrip feed cables and SIRS are used to engineer the band switching. Additionally, varactor diodes positioned at SIRS center points are used to tune frequency. By implementing high-impedance biasing lines, it is possible to achieve the separation between the DC source and RF signal. A 150 nH inductor RF choke is present on each bias line (Murata LQW18ANR25J80D27 is adopted in this design). To block DC at the input RF port, a 2.2 pF obstructing capacitor is also added to the feed line. So, without changing the number of ports or antenna elements, wideband-to-narrowband frequencies, and port adjustable antenna is achieved, which results in a smaller RF circuit.

4.9.3 Result and discussion

The proposed antenna is symmetrical in nature, and it comes in various reconfigurable states. S parameters of a single port in many reconfigurable states have been tested on a network analyzer. Figure 4.13 displays the measured and simulated S-parameters of the proposed antenna while maintaining match load termination on all other ports. The antenna functions in a 5G narrowband condition whenever diodes 1 and 2 are ON while maintaining all other diodes OFF condition. The wireless LAN narrowband state is achieved in a similar manner by setting 3 and 4 in the ON state while leaving all the other diodes OFF state. Figure 4.14 shows the computed and measured results both demonstrate that the suggested antenna has port isolation that is better than or equal to 16 dB in all three scenarios. While the WLAN state antenna is adjustable from 4.5 to 5.9 GHz, which also covers the n79 band of mid-band 5G, the 5G state antenna is controllable in the

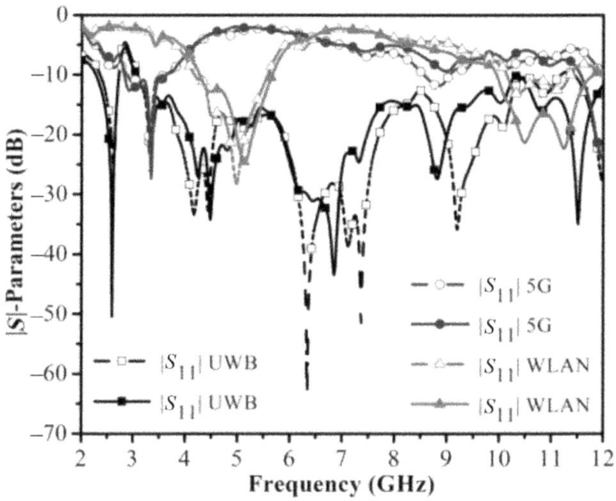

Figure 4.13 S_{11} of proposed MIMO antenna.

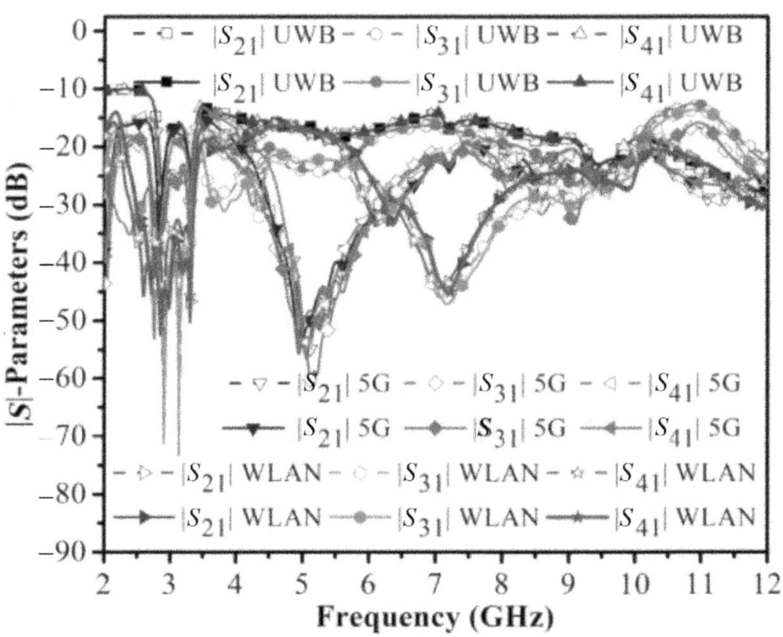

Figure 4.14 S parameters of proposed MIMO antenna.

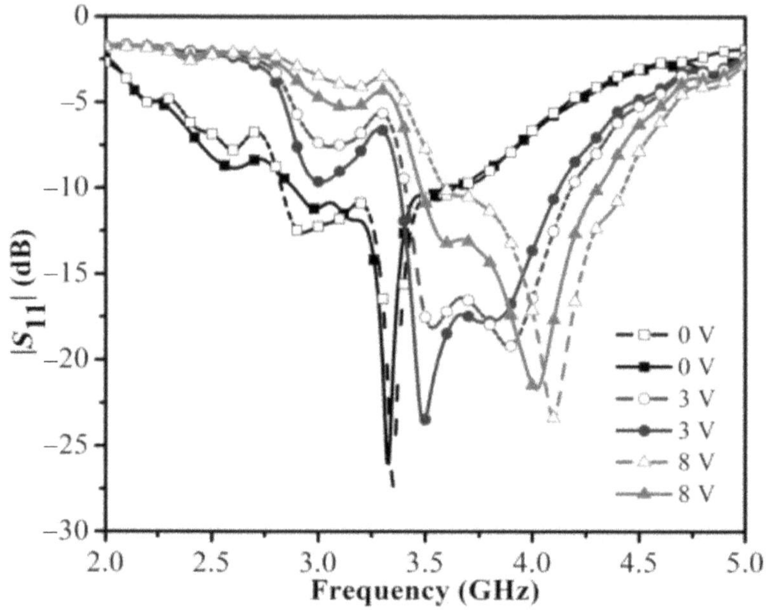

Figure 4.15 S_{11} of presented MIMO antenna with tunable pin diode.

region of 3–4.5 GHz, including the n77 and n78 5G band. Due to SIRS, the harmonics suppression of 5G and wireless LAN spectrum operation is nearly up to 10 GHz. In the anechoic room, the MIMO system radiation properties were examined, and the results were contrasted with simulations. Figure 4.15 presents the s parameters of the proposed MIMO antenna with a tunable pin diode.

Figure 4.16 displays radiation patterns that were both simulated and actually measured from two orthogonal ports in two principal planes. It also demonstrates that at lower frequencies, port 1 in the xz-plane and port 2 in the yz-plane produce almost figure-of-eight-shaped patterns, whereas port 1 in the yz-plane and port 2 in the xz-plane produce virtually omnidirectional patterns. Shape distortion from both ports in two planes at higher frequencies is caused by the presence of undesired higher-order modes. Figure 4.16 depicts identical radiation characteristics for mid-band 5G and WLAN states at 0 V bias as well as the wideband-narrowband adjustment at the feed line. The polarization diversity is explained by the radiation patterns from the two ports. Additionally, every radiation pattern is comparable to that of a traditional printed monopole antenna, guaranteeing good radiation characteristics. Figure 4.17 displays the proposed antenna's simulated and measured realized gains and radiation efficiency in all three states. In

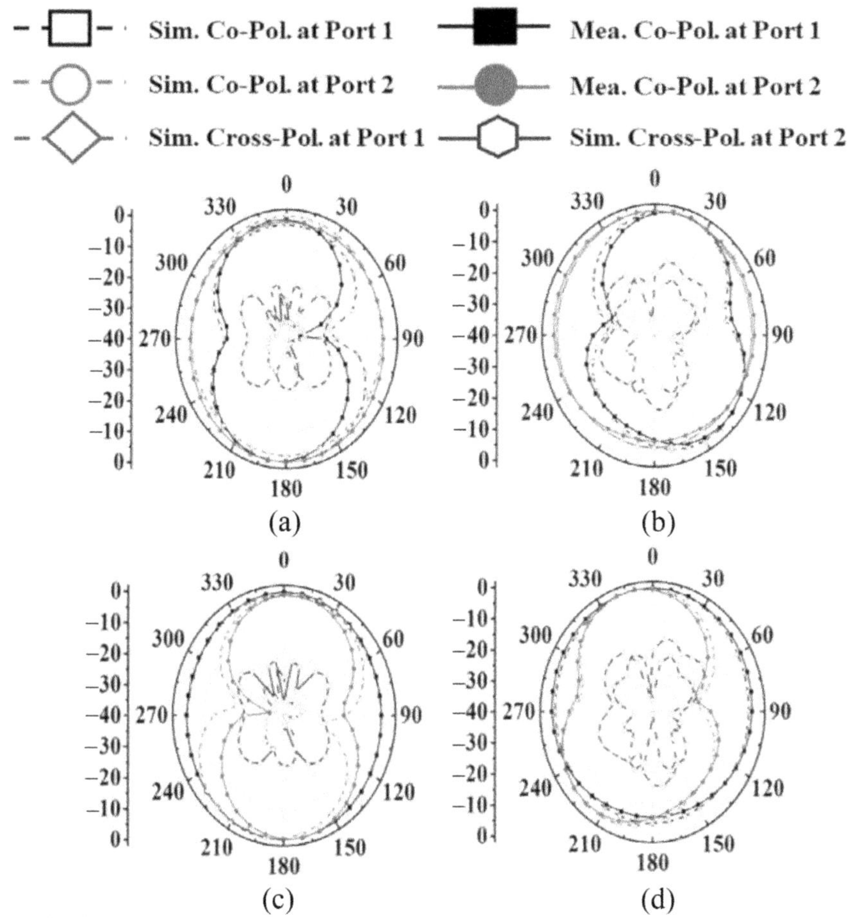

Figure 4.16 Radiation pattern of proposed MIMO antenna.

the UWB condition, the gain fluctuates from 3 to 4.6 dB. The peak realized gains in the narrowband range from 3 to 4 dB in the wireless LAN state and from 3.7 to 4 dB in the 5G state. At the greater frequencies, because of the formation of higher-order modes and substrate losses, the radiation effectiveness of an antenna fluctuates between 50% and 70%. While in 5G and WLAN states, the efficiency ranges from 60% to 70% in narrowband states. Wideband and narrowband states have a slight mismatch in gain and efficiency that is induced by losses from active components and stubs. The suggested antenna's MIMO diversity performance is examined in the UWB condition using the envelope correlation coefficient (ECC). According

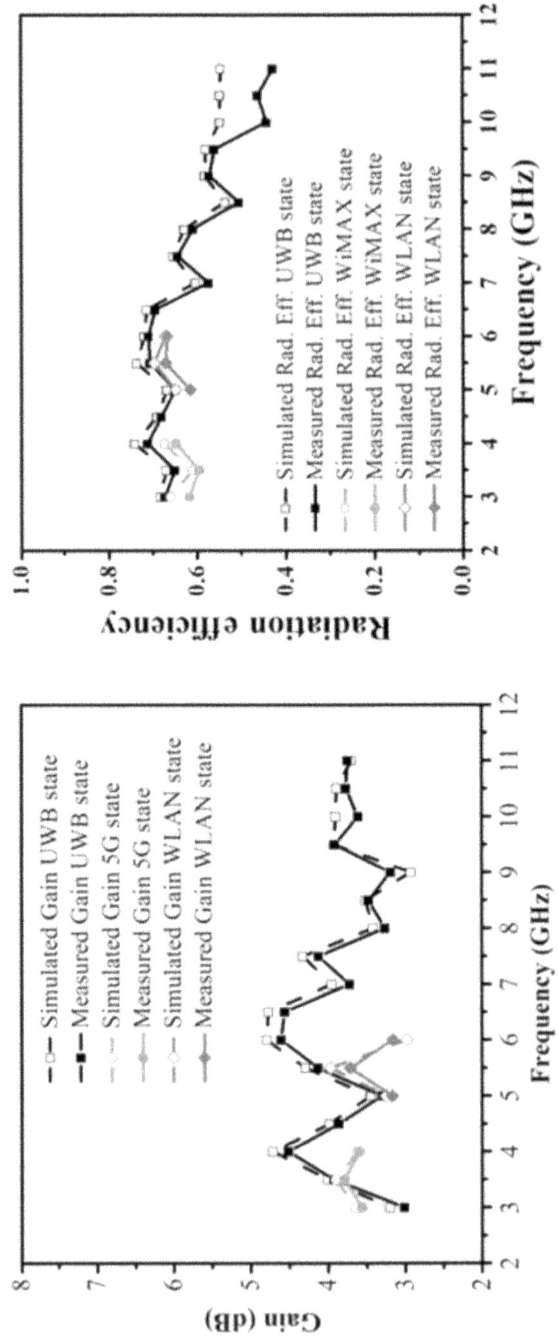

Figure 4.17 Gain and radiation efficiency of proposed MIMO antenna.

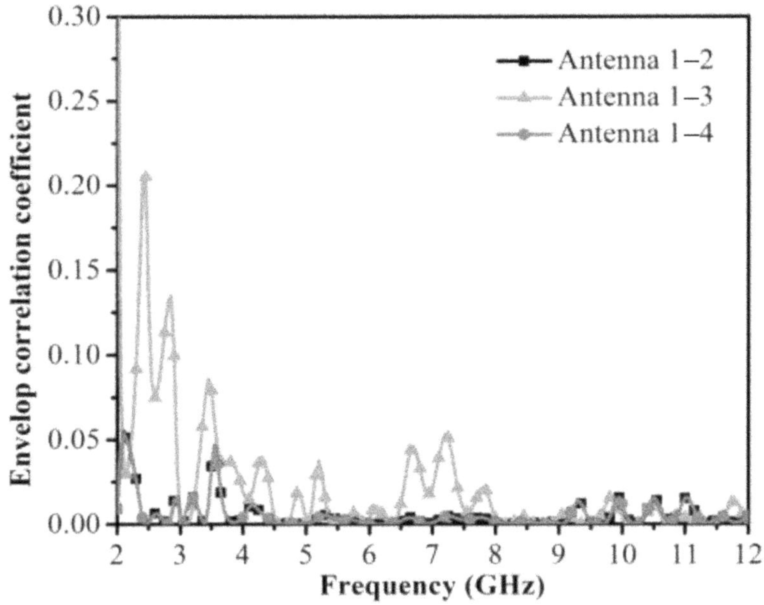

Figure 4.18 ECC of proposed MIMO antenna.

to Figure 4.18, the ECC is calculated for the isotropic scenario19 at port 1. The orthogonal polarization between two radiators that are adjacent to one another offers ECC from all ports below 0.15, which is much lower than the permitted limit of 0.5.19.

4.10 CONCLUSION

The chapter discusses the transformative impact of MIMO reconfigurable antennas on wireless communication systems. Combining MIMO's advantages with reconfigurable antennas, these technologies enhance spectral efficiency, data rates, and system performance. They offer adaptability to changing environments, enabling dynamic signal optimization. This synergy addresses challenges in modern wireless networks and paves the way for future advancements in 5G, IoT, and beyond. As the reconfigurable antennas are low in profile and can be made very compact these antennas are finding a wide variety of applications in IoT devices. The reconfigurability of the antenna can be achieved using different types of diodes. The demands of IoT-based devices have been met by frequency reconfigurability. The reconfigurable antenna can provide different wireless communication frequencies in either a 4-port MIMO configuration or two 2-port MIMO configurations.

The various diode states also provide multiband operation, where each port runs at a separate frequency. The proposed antenna provides port isolation that is better than 16 dB. By altering the reverse – biased voltage of varactor diodes, the narrowband states can be further tuned at various frequency bands. At fixed 2- and 4-port MIMO antenna arrangements, it has been shown that other antennas enable narrowband tuning. The proposed antenna, on the other hand, is small and has each port adjustable for both wideband and narrowband states that are employed for communication and sensing. The measured and simulated results agree fairly well. The reconfigurable frequency nature and real-time testing make reconfigurable antennas suitable for IoT and other wireless platform-based applications.

REFERENCES

Abu Tarboush, H. F., Nilavalan, R., Nasr, K. M., Cheung, S. W., Peter, T., Al-Raweshidy, H. S., & Budimir, D. (2011). Reconfigurable tri-band H-shaped antenna with frequency selectivity feature for compact wireless communication systems. *IET Microwaves, Antennas & Propagation*, *5*(14), 1675. https://doi.org/10.1049/iet-map.2010.0518

Ali, M. T., Kamarudin, M. R., Rahman, T. B. A., Sauleau, R., & Md Tan, M. N. (2009). Design of reconfigurable multiple elements microstrip rectangular linear array antenna. *Progress In Electromagnetics Research C*, *6*, 21–35. https://doi.org/10.2528/PIERC08122101

Bhatti, R. A., Choi, J.H., & Park, S.O. (2009). Quad-band MIMO antenna array for portable wireless communications terminals. *IEEE Antennas and Wireless Propagation Letters*, *8*, 129–132. https://doi.org/10.1109/LAWP.2008.2012274

Bolin, T., Derneryd, A., Kristensson, G., Plicanic, V., & Ying, Z. (2005). Two antenna receive diversity performance in indoor environment. *Electronics Letters*, 41(22), 1205–1206.

Chung, K., & Yoon, J. H. (2007). Integrated MIMO antenna with high isolation characteristic. *Electronics Letters*, *43*(4), 199. https://doi.org/10.1049/el:20070012

Cui, S., Liu, Y., Jiang, W., & Gong, S. X. (2011). Compact dual-band monopole antennas with high port isolation. *Electronics Letters*, *47*(10), 579. https://doi.org/10.1049/el.2010.3603

Diallo, A., Luxey, C., Le Thuc, P., Staraj, R., & Kossiavas, G. (2006). Study and reduction of the mutual coupling between two mobile phone PIFAs operating in the DCS1800 and UMTS bands. *IEEE Transactions on Antennas and Propagation*, *54*(11), 3063–3074.

Ding, Y., Du, Z., Gong, K., & Feng, Z. (2007). A novel dual-band printed diversity antenna for mobile terminals. *IEEE Transactions on Antennas and Propagation*, *55*(7), 2088–2096. https://doi.org/10.1109/TAP.2007.900249

Erdil, E., Topalli, K., Unlu, M., Civi, O. A., & Akin, T. (2007). Frequency tunable microstrip patch antenna using RF MEMS technology. *IEEE Transactions on Antennas and Propagation*, *55*(4), 1193–1196. https://doi.org/10.1109/TAP.2007.893426

Foschini, G. J., & Gans, M. J. (1998). On limits of wireless communications in a fading environment when using multiple antennas. *Wireless Personal Communications*, *6*(3), 311–335. https://doi.org/10.1023/A:1008889222784

Gallo, M., Antonino-Daviu, E., Ferrando-Bataller, M., Bozzetti, M., Molina-Garcia-Pardo, J. M., & Juan-Llacer, L. (2012). A broadband pattern diversity annular slot antenna. *IEEE Transactions on Antennas and Propagation*, *60*(3), 1596–1600. https://doi.org/10.1109/TAP.2011.2180314

Gao, P., He, S., Wei, X., Xu, Z., Wang, N., & Zheng, Y. (2014). Compact printed UWB diversity slot antenna with 5.5-GHz band-notched characteristics. *IEEE Antennas and Wireless Propagation Letters*, *13*, 376–379. https://doi.org/10.1109/LAWP.2014.2305772

Gokalp, N., & Civi, O. A. (2011). Beam-steerable meander line antenna using varactor diodes. *Microwave and Optical Technology Letters*, 53, 200–204.

Hamzah, S. A., Esa, M., Malik, N. N. N. A., & Ismail, M. K. H. (2012). Reconfigurable harmonic suppressed fractal dipole antenna. *2012 Asia Pacific Microwave Conference Proceedings*, 800–805. https://doi.org/10.1109/APMC.2012.6421740

Hong, S., Chung, K., Lee, J., Jung, S., Lee, S. S., & Choi, J. (2008). Design of a diversity antenna with stubs for UWB applications. *Microwave and Optical Technology Letters*, *50*(5), 1352–1356. https://doi.org/10.1002/mop.23389

Kaiser, T., Zheng, F., & Dimitrov, E. (2009). An overview of ultra-wide-band systems with MIMO. *Proceedings of the IEEE*, *97*(2), 285–312. https://doi.org/10.1109/JPROC.2008.2008784

Khan, M. S., Capobianco, A. D., Najam, A. I., Shoaib, I., Autizi, E., Shafique, M. S. (2014). Compact ultra-wideband diversity antenna with a floating parasitic digitated decoupling structure. *IET Microwaves, Antennas & Propagation*, *8*(10), 747–753.

Lai, M. I., Wu, T. Y., Hsieh, J. C., Wang, C. H., & Jeng, S. K. (2009). Design of reconfigurable antennas based on an L-shaped slot and PIN diodes for compact wireless devices. *IET Microwaves, Antennas & Propagation*, *3*(1), 47. https://doi.org/10.1049/iet-map:20080049

Lee, J. M., Kim, K. B., Ryu, H. K., & Woo, J. M. (2012). A compact ultrawideband MIMO antenna with WLAN band-rejected operation for mobile devices. *IEEE Antennas and Wireless Propagation Letters*, *11*, 990–993. https://doi.org/10.1109/LAWP.2012.2214431

Li, J. F., Chu, Q. X., Li, Z. H., & Xia, X. X. (2013). Compact dual band-notched UWB MIMO antenna with high isolation. *IEEE Transactions on Antennas and Propagation*, *61*(9), 4759–4766. https://doi.org/10.1109/TAP.2013.2267653

Li, Y., Zhang, Z., Zheng, J., Feng, Z., & Iskander, M. F. (2011). Experimental analysis of a wideband pattern diversity antenna with compact reconfigurable CPW-to-Slotline transition feed. *IEEE Transactions on Antennas and Propagation*, *59*(11), 4222–4228. https://doi.org/10.1109/TAP.2011.2164224

Liu, L., Zhao, H., See, T. S. P., & Chen, Z. N. (2006). A printed ultra-wideband diversity antenna. *2006 IEEE International Conference on Ultra-Wideband*, 351–356. https://doi.org/10.1109/ICU.2006.281575

Majid, H. A., Rahim, M. K. A., Hamid, M. R., & Ismail, M. F. (2012a). A compact frequency-reconfigurable narrowband microstrip slot antenna. *IEEE*

Antennas and Wireless Propagation Letters, *11*, 616–619. https://doi.org/10.1109/LAWP.2012.2202869

Majid, H. A., Rahim, M. K. A., Hamid, M. R., Ismail, M. F., & Malek, F. (2012b). Frequency reconfigurable wide to narrow band monopole with slotted ground plane antenna. *Journal of Electromagnetic Waves and Applications*, *26*(11–12), 1460–1469. https://doi.org/10.1080/09205071.2012.702536

Mak, A. C. K., Rowell, C. R., & Murch, R. D. (2008). Isolation enhancement between two closely packed antennas. *IEEE Transactions on Antennas and Propagation*, *56*(11), 3411–3419. https://doi.org/10.1109/TAP.2008.2005460

Mathur, R., & Dwari, S. (2021). Frequency and port reconfigurable MIMO antenna for UWB 5G WLAN band IoT applications. *International Journal of RF and Microwave Computer-Aided Engineering*, *31*(7). https://doi.org/10.1002/mmce.22692

Mohammad, S., Nezhad, A., Hassani, H. R., & Foudazi, A. (2013). A dual-band WLAN/UWB printed wide slot antenna for MIMO/diversity applications. *Microwave and Optical Technology Letters*, *55*(3), 461–465. https://doi.org/10.1002/mop.27391

Morabito, A. F., Isernia, T., & Di Donato, L. (2012). Optimal synthesis of phase-only reconfigurable linear sparse arrays having uniform-amplitude excitations. *Progress In Electromagnetics Research*, *124*, 405–423. https://doi.org/10.2528/PIER11112210

Pourziad, A., Nikmehr, S., & Veladi, H. (2013). A novel multi-state integrated RF MEMS switch for reconfigurable antennas applications. *Progress In Electromagnetics Research*, *139*, 389–406. https://doi.org/10.2528/PIER13012303

Qin, Y., Zhang, L., Mao, C. X., & Zhu, H. (2022). A compact wideband antenna with suppressed mutual coupling for 5G MIMO applications. *IEEE Antennas and Wireless Propagation Letters*, 22(4), 938–942. https://doi.org/10.1109/LAWP.2022.3229020

Ren, J., Hu, W., Yin, Y., & Fan, R. (2014). Compact printed MIMO antenna for UWB applications. *IEEE Antennas and Wireless Propagation Letters*, *13*, 1517–1520. https://doi.org/10.1109/LAWP.2014.2343454

Ren, Z., Li, W.-T., Xu, L., & Shi, X.-W. (2013). A compact frequency reconfigurable unequal U-slot antenna with a wide tunability range. *Progress In Electromagnetics Research Letters*, *39*, 9–16. https://doi.org/10.2528/PIERL13022508

Ruvio, G., Ammann, M. J., & Chen, Z. N. (2007). Wideband reconfigurable rolled planar monopole antenna. *IEEE Transactions on Antennas and Propagation*, *55*(6), 1760–1767. https://doi.org/10.1109/TAP.2007.898575

See, T. S. P., & Chen, Z. N. (2009). An ultrawideband diversity antenna. *IEEE Transactions on Antennas and Propagation*, *57*(6), 1597–1605. https://doi.org/10.1109/TAP.2009.2019908

Sorin, V., Artemenko, A., Sevastyanov, A., & Maslennikov, R. (n.d.). Compact Bandwidth-Optimized Two Element MIMO Antenna System for 2.5–2.7 GHz Band. In *Proceedings of the 5th European Conference on Antennas and Propagation (EUCAP)* (pp. 319–23). IEEE.

Wallace, J. W., Jensen, M. A., Swindlehurst, A. L., & Jeffs, B. D. (2003). Experimental characteriz ation of the MIMO wireless channel: data acquisition and analysis. *IEEE Transactions on Wireless Communications*, *2*(2), 335–343. https://doi.org/10.1109/TWC.2003.808975

Yang, C., Yao, Y., Yu, J., & Chen, X. (2012). Novel compact multiband MIMO antenna for mobile terminal. *International Journal of Antennas and Propagation*, *2012*, 1–9. https://doi.org/10.1155/2012/691681

Yang, X. S., Wang, B. Z., Wu, W., & Xiao, S. (2007). Yagi patch antenna with dual-band and pattern reconfigurable characteristics. *IEEE Antennas and Wireless Propagation Letters*, *6*, 168–171. https://doi.org/10.1109/LAWP.2007.895292

Yoon, H. K., Yoon, Y. J., Kim, H., & Lee, C. H. (2011). Flexible ultra-wideband polarisation diversity antenna with band-notch function. *IET Microwaves, Antennas & Propagation*, *5*(12), 1463–1470. https://doi.org/10.1049/iet-map.2010.0126

Zeng, Q., Yao, Y., Liu, S., Yu, J., Xie, P., & Chen, X. (2012). Tetraband small-size printed strip MIMO antenna for mobile handset application. *International Journal of Antennas and Propagation*, *2012*, 320582:1–320582:8.

Zhang, S., Huff, G. H., Cung, G., & Bernhard, J. T. (2005). Three variations of a pattern-reconfigurable microstrip parasitic array. *Microwave and Optical Technology Letters*, *45*, 369–372.

Zhang, S., Lau, B. K., Sunesson, A., & He, S. (2012a). Closely-packed UWB MIMO/diversity antenna with different patterns and polarizations for USB dongle applications. *IEEE Transactions on Antennas and Propagation*, *60*(9), 4372–4380. https://doi.org/10.1109/TAP.2012.2207049

Zhang, S., Ying, Z., Xiong, J., & He, S. (2012b). Reducing mutual coupling of MIMO antennas with parasitic elements for mobile terminals. *IEEE Transactions on Antennas and Propagation*, 60(2), 473–481.

Zhang, S., Ying, Z., Xiong, J., & He, S. (2009). Ultrawideband MIMO/diversity antennas with a tree-like structure to enhance wideband isolation. *IEEE Antennas and Wireless Propagation Letters*, *8*, 1279–1282. https://doi.org/10.1109/LAWP.2009.2037027

Chapter 5

Exploring the dynamics of microstrip antenna and radiation mechanism of dipole antennas

Sunaina Singh, Vartika Dahima, and Ranjan Mishra

5.1 INTRODUCTION

In the 1890s, there were only a few antennas in the world. These simple apparatuses were mostly used in investigations showing electromagnetic wave propagation. By World War II, radio and television reception had grown so commonplace that antenna use had completely changed people's lives. In the United States, there was around one antenna for every family, a growth rate that was comparable to that of the automobile sector in the same time frame. The ordinary individual today takes many antennas with them wherever they go in the early 21st century, largely due to mobile phones (which feature multiple antennas such as GPS, Wi-Fi, cellular, and near-field communication [NFC]). With the increasing integration of wireless communication devices into daily life, this noteworthy growth rate is unlikely to abate [1]. Furthermore, given the rapid expansion of radio frequency identification (RFID) devices, it is possible that there will soon be one antenna for every item on the planet (product, container, pet, banana, toy, CD, etc.). The number of antennas in use today would be dwarfed by this amount. A microstrip patch antenna is a form of antenna that is made up of a conductive patch that is flat and rectangular (or other shaped) and is mounted on a thin dielectric substrate. On the opposite side of the substrate is a conductive ground plane where the patch is mounted. As a result of its low profile, ease of production, and versatility, it is one of the antennas that is utilized most frequently in contemporary instances of wireless communication. An introduction to microstrip patch antennas, along with a list of popular applications for these antennas, is presented here. The versatility and compact design of microstrip patch antennas make them suitable for usage in a wide range of applications. Because of their compact size and the ease with which they may be integrated, these antennas are frequently utilized in wireless communication devices such as mobile phones, Wi-Fi routers, and other wireless devices. As a result of its low profile, microstrip patch antennas are well-suited for use in satellite communications, which

DOI: 10.1201/9781003470854-5

are characterized by the importance of aerodynamic design. They are utilized in radar applications, mainly when space limits are an issue, such as car radar for collision avoidance systems. Radar systems are also used in other applications. Microstrip patch antennas can be arranged in arrays to generate phased arrays or other complicated configurations, which enables beamforming and other advanced features. Antenna arrays are a type of antenna, Medical Devices: Because these antennas are compact and lightweight, they are suited for medical devices such as diagnostic equipment and wearable gadgets. Internet of Things (IoT) and Smart Devices: The proliferation of the IoT is significantly dependent on microstrip patch antennas for establishing connectivity in smart devices and sensors. This includes the Medical Implant (ISM) frequency range of 902–928 MHz, 950–958 MHz, 2360–2400 MHz, and 2400–2483 MHz. It also includes the ultra-wideband frequency range of 3.1–10 GHz, and the Medical Implant Communications Service (MICS) frequency range of 402–405 MHz.

Microstrip antennas are a particularly intriguing and new area of study in modern antenna technology because of their many benefits, including ease of manufacturing, low cost, lightweight, and smaller size and profile [2]. To meet the growing needs of several wireless devices with a single antenna, researchers are becoming more and more interested in creating an antenna with wideband and multiband properties [3]. It is employed in many different fields, including wireless fidelity, meteorological and military applications, and biomedical telemetry services. The components of a microstrip patch antenna are ground, dielectric substrate, and patch. The layer in between the patch and the ground is called the substrate. The patch is typically designed with copper and can take on a variety of designs, including square, rectangular, round, elliptical, and triangular [4]. In Figure 5.1 we define the structure of the microstrip antenna.

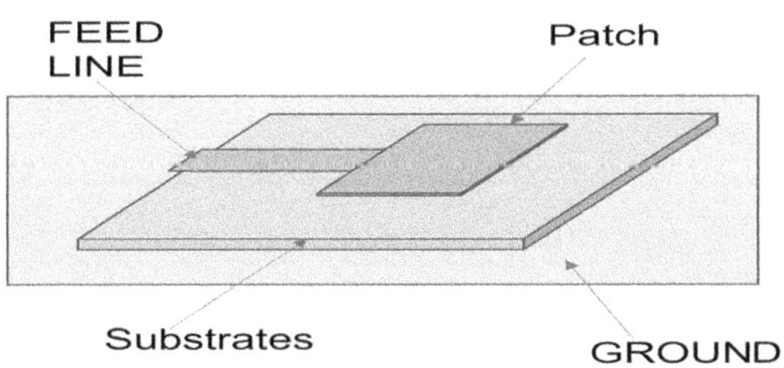

Figure 5.1 Structure of microstrip patch.

5.2 PATCH ANTENNA

Establishing the design specifications is the most important phase. The application is the primary factor that determines the design parameters. The system engineer typically provides antenna specifications after completing the system-level analysis and modeling of a full system, such as a radar system or a 5G network. When compared to the specifications of a satellite or radar antenna system, the specifications of a wireless mobile communication antenna are entirely different [1].

1. Calculate the width of the patch using line.

$$W = \frac{1}{f_r}\sqrt{\frac{2}{\varepsilon_r + 1}} \tag{5.1}$$

$$W = \frac{1}{2f_r\sqrt{\mu_0\varepsilon_0}}\sqrt{\frac{2}{\varepsilon_r + 1}} = \frac{v_0}{2f_r}\sqrt{\frac{2}{\varepsilon_r + 1}} \tag{5.2}$$

(Preferred for its ability to provide high radiation efficiency.) To ensure that the patch is only used in the TM010 mode, it is imperative that the width of the patch is smaller than the length of the patch (i.e., $W < L$). It is possible to utilize $W > L$ in practice if the patch is stimulated or driven in such a way that it does not excite other modes (for example, the TM001 mode is dominant when $W > L$ is employed). The value of W can be altered if the ratio of W to L is less than 2. This will prevent a loss in aperture efficiency.

3. Calculate

$$\varepsilon_{r,eff} = \frac{\varepsilon_r + 1}{2} + \frac{\varepsilon_r - 1}{2}\left[1 + 12\frac{h}{w}\right]^{-1/2} \tag{5.3}$$

4. Calculate

$$\Delta L = 0.412h\frac{\left(\varepsilon_{r,eff} + 0.3\right)\left(\frac{W}{h} + 0.624\right)}{\left(\varepsilon_{r,eff} + 0.3\right)\left(\frac{W}{h} + 0.624\right)} \tag{5.4}$$

5. Calculate the effective length and guided wavelength λ.

$$L_{eff} = \frac{C}{2f_r\sqrt{\varepsilon_{r,eff}}} \tag{5.5}$$

6. Calculate the patch length L (see below) and L/λ. Then, calculate and check that $W/L < 2$. If necessary, reduce the W calculated in step 2 to be slightly less than $2L$.

$$L_{eff} = L + 2\Delta L \tag{5.6}$$

$$L = \frac{1}{2f_r\sqrt{\epsilon_{reff}}\sqrt{\mu_0\epsilon_0}} - 2\Delta L$$

7. Calculate the characteristic impedance $Z_{c,ant}$ and admittance $Y_{c,ant}$ for the rectangular microstrip antenna.

$$Z_{c,ant} = \begin{cases} \dfrac{60}{\sqrt{\varepsilon_{r,eff}}} In\left[\dfrac{8h}{W} + \dfrac{W}{4h}\right] \\ \dfrac{n_0}{\sqrt{\varepsilon_{r,eff}}\left[\dfrac{W}{h} + 1.393 + 0.667In\left(\dfrac{W}{h} + 1.444\right)\right]} & \dfrac{W}{h} > 1 \end{cases} \tag{5.7}$$

5.2.1 Design of microstrip patch

The suggested geometric arrangement of the adaptable single-element antenna is designed for operation inside the sub-6 GHz band of frequency ranges. The planar patch configuration is readily comprehensible due to its utilization in the integration of smart handheld devices[5]. The chosen frequency for the wireless application is taken as 2.45 GHz. A microstrip line with a 50-ohm impedance is used to feed the antenna. The height of the inset feed point in relation to the patch's center is 30 mm. Table 5.1 contains the dimensions of the antenna in millimeters and Figure 5.2 shows the design of the microstrip inset-feed patch antenna.

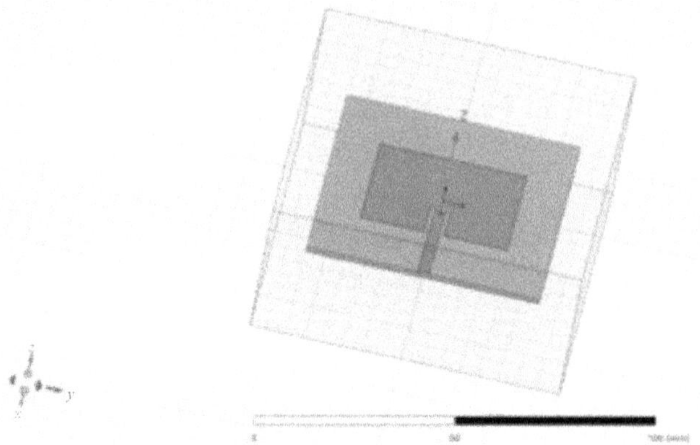

Figure 5.2 The design of microstrip patch antenna.

Table 5.1 Dimensions of designed antenna

S. No.	*Parameters*	*Dimensions (mm)*
1.	Substrate length	60
2.	Substrate width	60
3.	Height	1.6
4.	GND length	60
5.	GND width	60
6.	Patch length	29.4
7.	Patch width	38
8.	Feedline width	3
9.	Feedline length	30
10.	Cut length	9.5
11.	Cut width	5

The dimensions of the ground plane are kept at 60 mm × 60 mm. The FR-4 dielectric material is chosen with a height of 1.6 mm with a relative permittivity equal to 4.3. The overall dimensions of the antenna that was designed were 60 mm × 60 mm × 1.6 mm.

In this structure has been found to have a resonance with a return loss value of –29 dB and a restricted bandwidth centered at 2.21 GHz using S_{11} parameter analysis. We use transmission line equations to find the patch dimensions. These dimensions are next optimized for the desired results via parametric analysis (Figure 5.3).

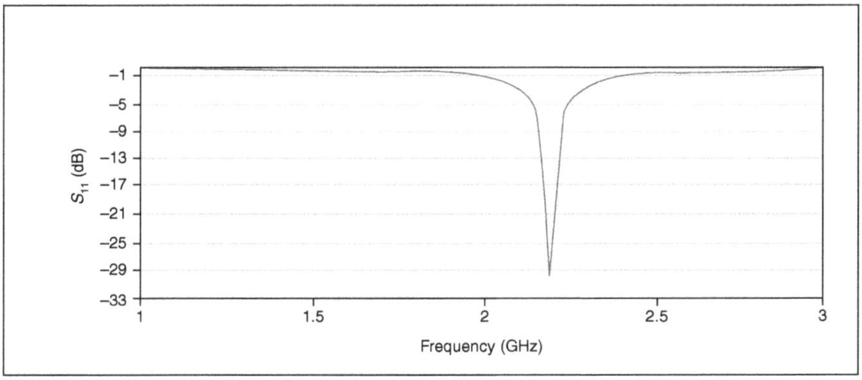

Figure 5.3 The S11 plot of antenna.

5.2.2 Parametric analysis

Within the context of antennas, parametric analysis refers to the process of systematically altering various factors that determine the antenna's shape, material qualities, and operating circumstances to gain an understanding of the effects that these parameters have on the antenna's performance characteristics. The radiation pattern, impedance matching, bandwidth, efficiency, and gain are some of the features that are often included in this category. The use of parametric analysis is essential for optimizing antenna designs to fulfill requirements and achieve desired levels of performance.

5.2.2.1 Variation of the length of the patch

The patch antenna has these fundamental dimensions according to its design. The process of parametric analysis entails adjusting these dimensions to gain an understanding of the influence that they have on the resonance frequency, bandwidth, and radiation characteristics of the antenna. Furthermore, in addition to the traditional rectangular patch, alternative shapes such as circular, triangular, or elliptical patches can be investigated to determine how they influence the performance of the antenna. Figure 5.4 demonstrates patch length variation from 26 to 29 mm. In the graph that in the parametric analysis is taken with a step size of 1 mm, bandwidth decreases significantly, with a resonance of 2.21 GHz and |S11| ranging from –30 dB to –35 dB; by increasing the patch length beyond 1 mm, the mode of excitation is offset feed resonance shifts to the lower side at 2.5 GHz while bandwidth decreases.

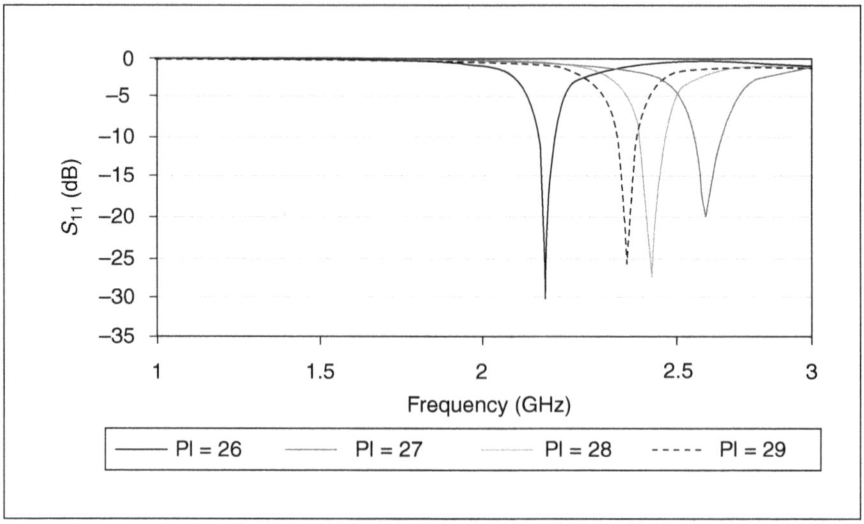

Figure 5.4 Parametric analysis of length of the patch antenna.

5.2.2.2 Variation of the length of the ground plane

We performed a parametric analysis on a ground antenna conducting a systematic investigation into the influence of several parameters on the antenna's operation. Applications such as radio transmission, broadcasting, and RFID systems extensively use ground antennas, often referred to as ground plane antennas or monopole antennas. The antenna's performance is substantially impacted by the ground plane located beneath it. When conducting parametric analysis, it is necessary to adjust the ground plane's dimensions, shape, and material qualities to understand how these factors influence antenna attributes, such as radiation pattern, gain, and impedance. Figure 5.5 demonstrates patch length variation from 67 to 70 mm. The graph indicates that for lengths under 1 mm, bandwidth decreases significantly, with a resonance of 2.21 GHz and |S11| ranging from −30 to −35 dB; by increasing the patch length beyond we are taking step size 1 mm, resonance shifts to the lower side at 2.21 GHz while bandwidth decreases. S11 is −30.11 dB.

5.2.2.3 Two- (2D) and three-dimensional (3D) radiation pattern

A plot of the radiation pattern of an antenna that operates at 2.21 GHz is shown in Figure 5.6(a). Using this picture, we can better comprehend the way the antenna emits energy in a variety of directions. The pattern has a few unique lobes and nulls, which can be used to determine the directivity

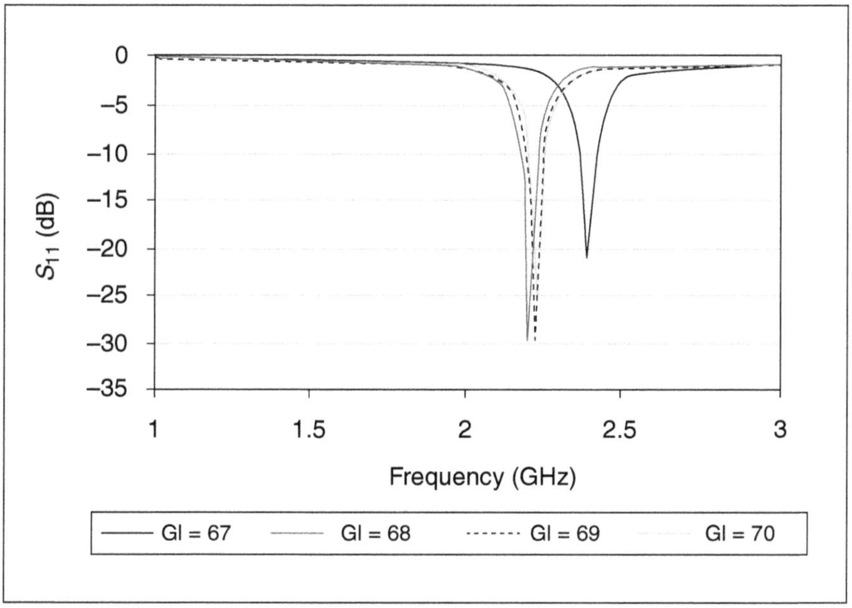

Figure 5.5 Parametric analysis of ground length.

and coverage of the antenna. A graphical representation of the 3D gain plots of the microstrip patch antenna may be found in Figure 5.6(b). This antenna can achieve a maximum gain of 2.7 dB.

5.2.2.4 E-field and H-field

When it comes to electromagnetic waves, the E-field and the H-field are orthogonal to one another and move in a path that is vertical to the way in which the wave is moving. According to the wave equation, the magnitudes of the E-field and the H-field are connected by the intrinsic impedance of the medium through which the wave is propagating. As they travel through space, changes in the E-field cause changes in the H-field, and changes in the H-field cause changes in the E-field. As a fundamental component of electromagnetic waves, the electric field is an essential component. When discussing antennas, the term "E-field" refers to the intensity of the electric field that is present in the area that is surrounding the antenna. When it comes to transmitting antennas, the E-field is the one that is accountable for inducing voltage in surrounding conductors. This includes reception antennas and circuits in communication equipment, respectively. Interactions between the E-field and the structure of the antenna in receiving antennas result in the generation of a voltage that may be detected and processed by electronic

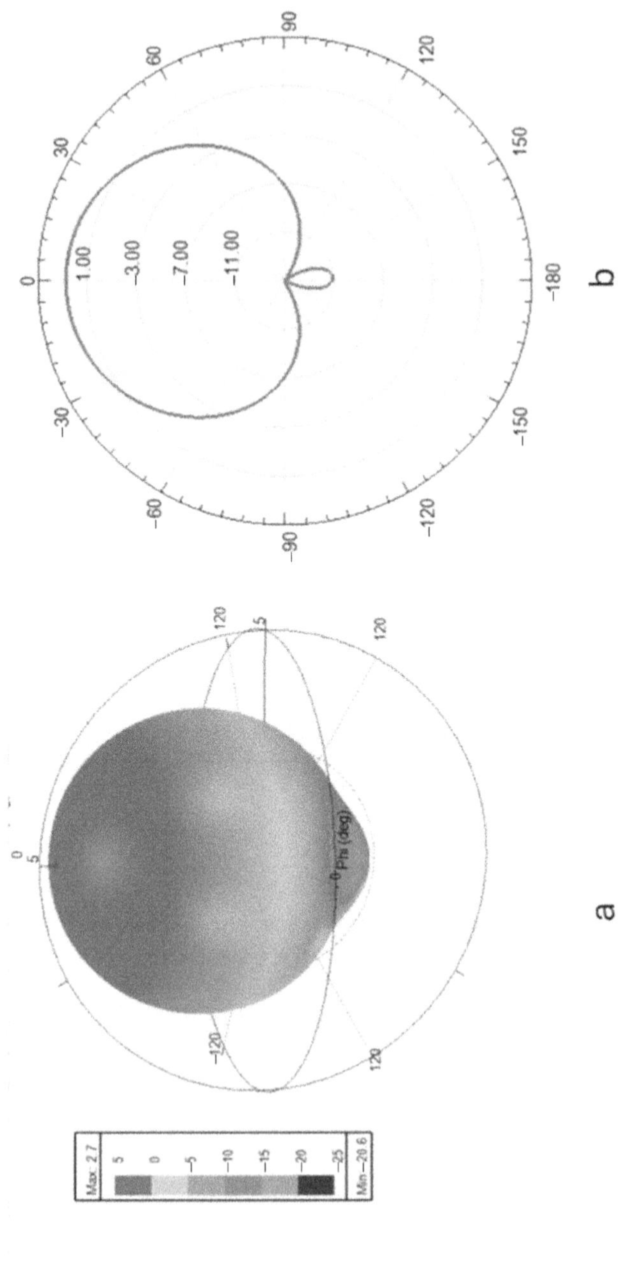

Figure 5.6 3D and 2D radiation pattern.

circuits that are attached to the antenna. An antenna's radiation pattern, polarization, and impedance properties are ultimately determined by the distribution of the electromagnetic field (E-field) around the antenna. It is common practice to depict the strength of an electromagnetic field using vector arrows that point in the direction of the field and are scaled to indicate the intensity of the field at various positions in space.

The electric and magnetic fields are the two components that make up electromagnetic waves. Regarding antennas, the H-field represents the strength of the magnetic field in the area surrounding the antenna. Inducing current in surrounding conductors, such as receiver antennas or circuits in communication equipment, is the responsibility of the H-field due to its ability to induce current. The radiation pattern, polarization, and impedance of the antenna are all affected by the interaction between the H-field and the structure of the antenna because of this interaction. The strength of the H-field can also be represented by vector arrows that point in the direction of the field and are scaled to indicate the intensity of the field at various positions in space. Figure 5.7 shows the E-field and H-field of the antenna.

5.3 DIPOLE ANTENNA

The dipole antenna is a straightforward antenna that is frequently utilized in a variety of applications, including radio broadcasting, wireless communications, and others. A feed point is in the middle of the structure, which is made up of a straight conductor or a pair of conductors that are normally placed in a symmetrical pattern. The most fundamental type of dipole antenna is a straight metal rod or wire divided into two equal portions. At the center of the antenna is a feed point connected to a transmission line. According to the operating frequency, the length of a dipole antenna is normally equal to half of the wavelength ($\lambda/2$) at that frequency. For example, the wavelength is roughly one meter while operating at a frequency of 300 MHz; hence, the length of a half-wave dipole would be nearly half two meters. The dipole antenna can be found in several different varieties, such as the half-wave dipole, the quarter-wave dipole, the folded dipole, and dipoles with additional components such as reflectors and directors. With a distinctive "figure-eight" form, the radiation pattern of a dipole antenna is typically omnidirectional in the horizontal plane. There is no discernible direction of radiation. Broadcasting and other applications that require uniform radiation can benefit from this pattern, which makes it advantageous for those applications. The orientation of dipole antennas determines whether they are polarized horizontally or vertically. Polarization can be either horizontal or vertical. The waves radiated by horizontal dipoles are horizontally polarized, whereas those emitted by vertical dipoles exhibit vertical polarization. Considering their ease of use and the fact that they emit radiation in all directions, dipole antennas are extensively used in

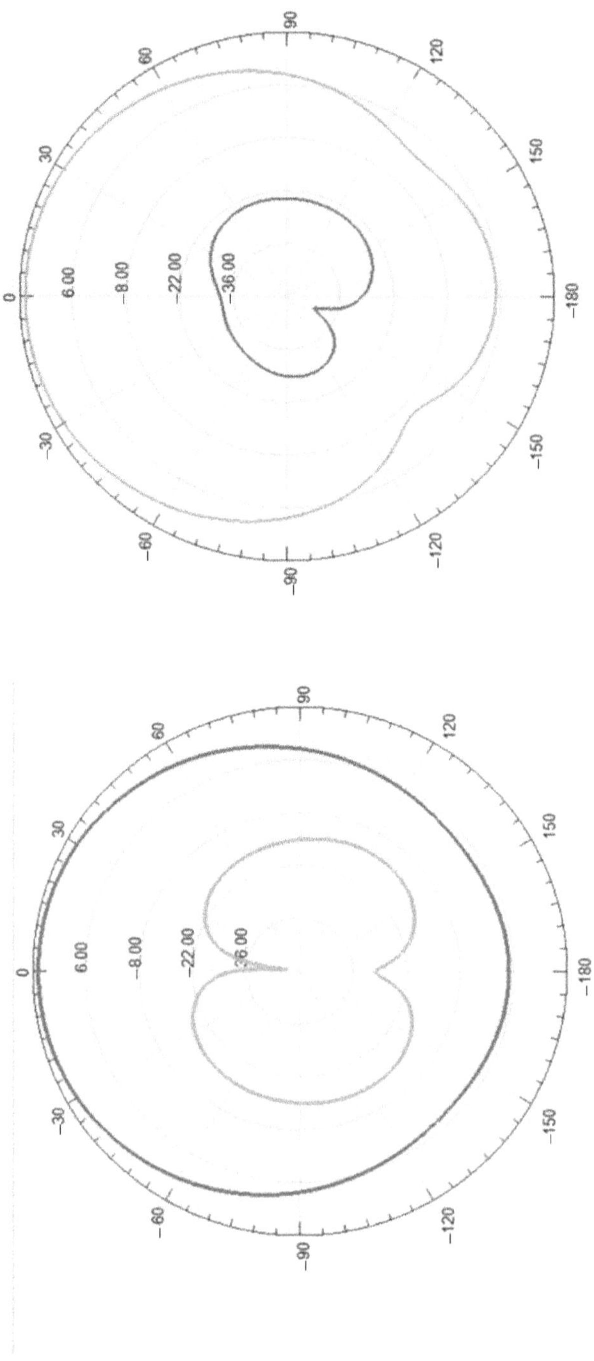

Figure 5.7 E-field and H-field.

radio and television broadcasting. They are ubiquitous in wireless communication systems, such as Bluetooth, Wi-Fi, and cellular networks, among others. Because they are simple to construct and have a wide frequency range, dipoles are trendy among amateur radio operators. Because their radiation patterns are predictable, dipoles are utilized as reference antennas in applications that involve measurement and testing. A dipole antenna is a straightforward one that is simple to construct and has predictable radiation patterns. In addition to being adaptable to various applications, it can be built to accommodate a wide range of frequencies. Dipoles generally resonate at a particular frequency, restricting the bandwidth they may operate within. Additionally, to connect to unbalanced coaxial transmission lines, they might need to implement a balanced-to-unbalanced transformer, commonly known as a balun.

5.4 RADIATION MECHANISM OF DIPOLE ANTENNA

The use of an alternating voltage to excite a transmitting antenna makes it possible for the first motion to be initiated by the balanced movement of charges within the antenna. This is because the antenna is used to send signals. The motion that you see here is caused by the antenna being energized. A resonant vibration is generated through the utilization of the energy that is supplied. An electric and magnetic field is generated because of the sudden shift in charge levels. Around the same time, the charges encircling the antenna start to move, and additional charges become disengaged from the antenna and start moving on their own. A disturbance that begins at the antenna and continues out into space is said to have originated there. The electric and magnetic fields generated because of this operation are perpendicular to one another throughout their whole life. The transmission of electromagnetic energy through space results in a reduction in its total amount. It is reference [6]. It is of theoretical importance to consider the concept of an alternating current (AC) element, which is denoted by $I\ dl$ cos θt. However, the theory created for this can be applied to antennas used in practice. When an alternating voltage is given to a transmitting antenna, the idea of retarded vector, which refers to the balanced motion of the antenna's charges, causes motion to be initiated within the antenna. The production of electric and magnetic fields is a consequence of sudden shifts in charge. In activating the charges surrounding the antenna, the extra charges are first separated from the antenna and then set in motion after being removed from the antenna. During the process of the disturbance being sent from the antenna, it is distributed into space. There are magnetic and electric fields perpendicular to one another that are produced in this manner. Electromagnetic waves are devoid of national boundaries 7]. When electromagnetic energy is propagated, the magnetic potential decreases. A is a beneficial tool to derive radiation fields of antenna elements, including the elements [8].

In AC element is positioned at the origin of a spherical coordinate system. It is the magnetic potential of the retarded vector, $\mathbf{A}(r, t)$ is given by.

$$A(r,t) = \frac{\mu}{4\pi}\int \frac{J(r,t - l v_0)}{r} dv \tag{5.8}$$

A is also z-directed considering that the element is z-directed. $\frac{r}{v_0}$ is the delay time.

To simplify the volume integral in Equation (5.1), it is possible to take the integration over the cross-sectional area of the element and an integration along its length into consideration. We are aware of

$$\oint_s^0 J ds = I \tag{5.9}$$

and

$$\oint_0^{dl} I \, dL = Idl \tag{5.10}$$

r denotes the distance between the receiving point and the center of the dipole. Idl represents the magnetic moment of the diploe the length of the dipole is $2L$.

$$A = A_z a_z = \frac{\mu_0}{4\pi}\frac{Idl \cos\omega(t - r/v_0)}{r} a_z \tag{5.11}$$

The z-component is the only one that A possesses, and $A_z = 0$ and $A_y = 0$.

When we shift the components of Cartesian coordinates to those of spherical coordinates, we get

$$\left.\begin{aligned} A_r &= A_z cos\theta \\ A_\theta &= -A_z sin\theta \\ A_\varnothing &= 0 \end{aligned}\right\} \tag{5.12}$$

But we know $H = \frac{1}{\mu_0}\nabla \times A$.

Through mathematical operations such as gradient and curl, the field vectors (electric and magnetic) and the potential vectors (scalar and vector potentials) are inextricably related to one another. The interactions between them offer essential insights into the generation and behavior of electromagnetic fields, which serve as the foundation for comprehending and resolving complicated issues in the field of electromagnetism.

$$\text{So}, H_r = \frac{1}{\mu_0}(\nabla \times A)_r$$

$$H_\theta = \frac{1}{\mu_0}(\nabla \times A)_\theta$$

$$H_\varnothing = \frac{1}{\mu_0}(\nabla \times A)_\varnothing$$

$$H_r = \frac{1}{\mu_0 r \sin\theta}\left[\frac{\partial}{\partial\theta}\left(\sin\theta A_\varnothing\right) - \frac{\partial A_\varnothing}{\partial\varnothing}\right]$$

As $A_\varnothing = 0$ and $A_\theta \neq f(\varnothing), H_r = 0$.

Similarly, $H_\theta = \frac{1}{\mu_0}\left[\frac{1}{r\sin\theta}\frac{\partial A_r}{\partial\varnothing} - \frac{1}{r}\frac{\partial\left(rA_\varnothing\right)}{\partial t}\right] = 0$ [as $A_r \neq f(\varnothing)$]

$$H_\theta = \frac{1}{\mu_0 r}\left[\frac{\partial}{\partial t}\left(rA_\theta - \frac{\partial A_r}{\partial t}\right)\right] \tag{5.13}$$

$$A_\theta = -A_z \sin\theta$$

$$= -\frac{\mu_0}{4\pi}\frac{Idl\cos\omega\left(t - r/v_0\right)}{r}\sin\theta \tag{5.14}$$

$$A_r = A_z \cos\theta$$

$$= -\frac{\mu_0}{4\pi}\frac{Idl\cos\omega\left(t - r/v_0\right)}{r}\cos\theta \tag{5.15}$$

Substituting Equations (5.7) and (5.8) with Equation (5.6) and simplifying, we get

$$H_{\varnothing} = \frac{Idl\sin\theta}{4\pi}\left[\frac{-\omega\sin\omega\left(t-\frac{r}{v_0}\right)}{rv_0} \div - \frac{\cos\omega\left(t-\frac{r}{v_0}\right)}{r^2}\right]$$

From Maxwell's first equation, we have

$$\nabla \times H = D = \epsilon_0 E$$

$$E = \frac{1}{\epsilon_0}\int(\nabla \times H)dt$$

In classical electromagnetism, Maxwell's equations are considered foundational principles. These equations describe the behavior of electric and magnetic fields and their interactions with electric charges and currents. These equations combine the areas of electricity, magnetism, and optics, and they also describe how these fields travel through space, which ultimately led to the development of the idea of electromagnetic waves[9]. Maxwell's equations and the correlations they establish between electric fields, magnetic fields, electric charges, and currents will be the subject of our investigation.

From Equations (5.6) and (5.9), we get E_r, E_θ, and $E_{\varnothing}$.
But $E_{\varnothing} = 0$ [as $H = H_{\varnothing}a_{\varnothing}$].

On simplifications,

$$E_\theta = \frac{Idl\sin\theta}{4\pi\epsilon_0}\left[-\frac{\omega\sin\omega t_d}{rv_\theta^2} + \frac{\cos\omega t_d}{r^2 v_0} + \frac{\sin\omega t_d}{\omega r^2}\right] \tag{5.16}$$

$$E_r = \frac{2\,Idl\sin\theta}{4\pi\epsilon_0}\left[\frac{\cos\omega t_d}{r^2 v_0} + \frac{\sin\omega t_d}{\omega r^2}\right] \tag{5.17}$$

$$t_d = \left(t - {}^{r}\!/_{v_0}\right) \tag{5.18}$$

The field components resulting from an AC element are.

$$H_{\varnothing} = \frac{Idl\sin\theta}{4\pi}\left[\frac{-\omega\sin\omega t_d}{rv_0} + \frac{\cos\omega t_d}{r^2}\right]$$

$$E_{\theta} = \frac{Idl\sin\theta}{4\pi\epsilon_0}\left[-\frac{\omega\sin\omega t_d}{rv_{\theta}^2} + \frac{\cos\omega t_d}{r^2 v_0} + \frac{\sin\omega t_d}{\omega r^3}\right] \quad (5.19)$$

$$E_r = \frac{2\ Idl\sin\theta}{4\pi\epsilon_0}\left[\frac{\cos\omega t_d}{r^2 v_0} + \frac{\sin\omega t_d}{\omega r^2}\right]$$

$$H_r = 0\ , H_{\theta} = 0,\ E_{\varnothing} = 0$$

5.5 POWER RADIATION AND RADIATION RESISTANCE OF AN ELECTRIC DIPOLE

Radiation is the term used to represent the emission or reception of wave front at the antenna, specifying its strength. In any illustration, the sketch drawn to represent the radiation of an antenna is its radiation pattern. One can simply understand the function and directivity of an antenna by having a look at its radiation pattern [10].

The methods involved in deriving the expression for radiated power are as follows:

Acquire field components. Using the Poynting vector, get the expression for the radiated power. Find the average power that is radiated. Determine the total power emitted by

$$P_T = \oint P_{av} ds$$

Identify protection against radiation.

Vector Poynting can be determined as under:

$$P = E \times H \text{ watts/m}^2$$

$$\text{That is, } P_{\theta} = -E_r H_{\varnothing} \quad (5.20)$$

From Equations (5.12) and (5.13), we have

$$P_\theta = \left[\frac{2Idl\cos\theta}{4\pi\epsilon_0}\left(\frac{\cos\omega t_d}{r^2 v_0} + \frac{\sin\omega t_d}{\omega r^3}\right)\right]$$

$$\times\left[\frac{2Idl\cos\theta}{4\pi}\left(-\frac{\omega\sin\omega t_d}{r v_0} + \frac{\cos\omega t_d}{r^2}\right)\right] \tag{5.21}$$

$$= \frac{I^2 dl^2 \sin\theta}{16\pi^2\epsilon_0}\left[\frac{\sin^2\omega t_d}{r^4 v_0} - \frac{\cos^2\omega t_d}{r^4 v_0} - \frac{\sin\omega t_d \cos\omega t_d}{2\omega r^5}\right.$$

$$\left. + \frac{\omega\sin\omega t_d \cos\omega t_d}{r^3 v_0^2}\right] \tag{5.22}$$

But

$$\left.\begin{aligned} &2\sin\theta = \sin 2\theta \\ &\sin\omega t_d \cos\omega t_d = \frac{\sin 2\omega t_d}{2} \\ &\sin\omega t_d - \cos^2\omega t_d = \cos 2\omega t_d \end{aligned}\right\} \tag{5.23}$$

Using these identities equation

$$P_\theta = \frac{I^2 dl^2 \sin\theta}{16\pi^2\epsilon_0}\left[-\frac{\cos 2\omega t_d}{r^4 v_0} + \frac{\sin 2\omega t_d}{2\omega r^5} + \frac{\omega\sin 2\omega t_d}{2 r^3 v_0^2}\right] \tag{5.24}$$

P_θ The power flow in θ--direction is represented instantaneously in Equation (5.17). However, the mean value of cos 2 ωt_d or sin 2 ωt_d cycle is zero. Where, $(P_\theta)_{av} = 0$ at any value of r. This means that power in θ-direction surges back and forth.

Similarly,

$$P_r = E_\theta H_\varnothing \tag{5.25}$$

From Equations (5.12) and (5.18), we get

$$P_\theta = \frac{I^2 dl^2 \sin\theta}{16\pi^2\epsilon_0}\left[\frac{\sin\omega t_d \cos\omega t_d}{2\omega r^5} + \frac{\cos\omega t_d}{r^4 v_0} + \frac{\omega \sin\omega t_d \cos\omega t_d}{r^3 v_0^2} - \frac{\sin^2\omega t_d}{r^4 v_0} - \frac{\omega\sin\omega t_d \cos\omega t_d}{r^3 v_0^2} + \frac{\omega^2 \sin^2\omega t_d}{r^2 v_0^3}\right] \tag{5.26}$$

Using identities of equation (5.26),

$$P_\theta = \frac{I^2 dl^2 \sin\theta}{16\pi^2\epsilon_0}\left[\frac{\sin^2\omega t_d}{2\omega r^5} + \frac{\cos 2\omega t_d}{r^4 v_0} - \frac{\omega \sin 2\omega t_d}{r^3 v_0^2} + \frac{\omega^2\left(1-\cos 2\omega t_d\right)}{2r^2 v_0^3}\right] \tag{5.27}$$

From this equation, we have the average value of P_r which is

$$P_{r(av)} = \frac{\omega^2 I^2 dl^2 \sin^2\theta}{32\pi^2 r^2 \epsilon_0 v_0^3}$$

$$P_{r(av)} = \frac{\eta_0}{2}\left(\frac{\omega I dl \sin\theta}{4\pi\epsilon_0}\right)^2 \text{ watts/m}^2 \tag{5.28}$$

where $\eta_0 = \sqrt{\frac{\mu_0}{\epsilon_0}} = 120\pi\Omega$.

The total power radiated.

$$P_T = \oint P_{r(av)} ds \tag{5.29}$$

$P_{r(av)}$ is independent of ∅ and consequently the element of area, d_s On the spherical shell is given by:

$$d_s = 2\pi r^2 \sin\theta d_\theta \tag{5.30}$$

Here, θ varies between 0 and π.

Now Equation (5.24) becomes

$$P_{r(av)} = \int_0^{\pi} \frac{\eta_0}{2} \left[\frac{\omega I dl \sin\theta}{4\pi r v_0} \right] 2\pi r^2 \sin\theta d_\theta$$

$$= \frac{\eta_0 \omega^r I^r dl^r}{16\pi v_0 r} \int_0^{\pi} \sin^3 \theta d_\theta$$

$$= \left(-\cos\theta + \frac{1}{3}\cos^3\theta \right)_0^{\pi} = \frac{4}{3}$$

So,

$$P_T = \frac{\eta_0 \omega^2 I^2 dl^2}{12\pi v_0^2} \tag{5.33}$$

I represent the peak value of current, as

$$I = \sqrt{2 I_{eff}}$$

$$I^2 = 2 I_{eff}^2$$

Thus Equation (5.24) becomes

$$P_T = \frac{\eta_0 \omega^2 I^2 dl^2}{6\pi v_0^2}$$

or

$$P_T = 80\pi^2 \left(\frac{dl}{\lambda} \right)^2 I_{eff}^2 \text{ watts} \tag{5.34}$$

The following can be expressed as $P = I^2 \cdot R$. Thus, the coefficients of I_{eff}^2

Radiation resistance of Hertzian dipole

$$R_r = 80\pi^2\left(\frac{dl}{\lambda}\right)^2 \Omega \tag{5.35}$$

These are the current elements' field components from Equation (5.12).

$$H_r = \frac{Idl\sin\theta}{4\pi\epsilon_0}\left[-\frac{\omega\sin\omega t_d}{r^2 v_0}+\frac{\cos\omega t_d}{r^2}\right]$$

$$E_\theta = \frac{Idl\sin\theta}{4\pi\epsilon_0}\left[-\frac{\omega\sin\omega t_d}{r v_0^2}+\frac{\cos\omega t_d}{r^2 v_0}+\frac{\sin\omega t_d}{\omega r^3}\right] \tag{5.36}$$

$$E_r = \frac{2\,Idl\sin\theta}{4\pi\epsilon_0}\left[\frac{\cos\omega t_d}{r^2 v_0}+\frac{\sin\omega t_d}{\omega r^3}\right]$$

$H_\varnothing$ field consists of $\frac{1}{r}$ and $\frac{1}{r^2}$ terms. $\frac{1}{r^2}$ term dominates over $\frac{1}{r}$ approximately at a place near the existing elements. For tiny values of r, $\frac{1}{r^2}$ term is called induction field. Conversely, however, $\frac{1}{r}$ term dominates over $\frac{1}{r^2}$ term when r is large. Here $\frac{1}{r}$ radiation field, far field, or distant field is the phrase used. The expression for E_θ consists of three terms: $\frac{1}{r}$, $\frac{1}{r^2}, \frac{1}{r^3}$ and the expression of E_r consists of $\frac{1}{r^2}$ and $\frac{1}{r^3}$ term. This $\frac{1}{r^3}$ term is called electrostatics field. $\frac{1}{r}$ radiation field is the phrase used in the H and I fields. $\frac{1}{r^2}$ the term is referred to as the induction field. $\frac{1}{r^3}$ the term is known as the electrostatic field. If the amplitudes of the induction field and the radiation field are equal, then according to Equations (5.28), we have $\frac{Idl\omega\sin\theta}{4\pi v_0} = \frac{Idl\sin\theta}{4\pi r^2}$

$$\frac{\omega}{rv_0^2} = \frac{1}{r^2} or\ r = \frac{v_0}{\omega} = \frac{\lambda}{2\pi} \approx \frac{\lambda}{6.0} \tag{5.37}$$

5.6 HERTZIAN DIPOLE

Real-world examples of tiny current elements are nonexistent, such as the Hertzian dipole. Alternatively, the Hertzian dipole is a short linear antenna that radiates and is thought to have a continuous current running along its length.

Radiation resistance and radiated power are both determined by the following:

As the Hertzian dipole and AC elements are nearly equivalent.

$$P_T = 80\pi^2 \left(\frac{dl}{\lambda}\right)^2 I_{eff}^2 \text{watts}$$

$$R_r = 80\pi^2 \left(\frac{dl}{\lambda}\right)^2 \Omega \tag{5.38}$$

5.7 LINEAR ANTENNAS HAVE A VARIETY OF DIVERSE CURRENT DISTRIBUTIONS

The potential current scatterings are:

1. Valid for a constant current in a Hertzian dipole along its length.
2. Triangle current distribution applicable to monopoles and short dipoles.

   For a triangular current distribution

$$R_r \left(short\ dipole, l < \frac{\lambda}{4}\right) \approx 20\pi \left(\frac{l}{\lambda}\right)^2 \Omega \tag{5.39}$$

$$R_r \left(short monopole, l < \frac{\lambda}{8}\right) \approx 10\pi \left(\frac{l}{\lambda}\right)^2 \Omega$$

3. Sinusoidal current distribution is applicable in a half-wave dipole.
4. The current circulation exactly as it is right now can be determined using the method of moments. However, this technique is beyond the scope of this book.

5.8 RADIATION FROM HALF-WAVE DIPOLE

The half-wave dipole as shown in Figure 5.8, with a length of λ/2, is a widely utilized antenna due to its radiation resistance of 73Ω at resonance. This value is near the characteristic impedances of 50Ω or 75Ω often seen in feeding transmission lines. Therefore, the process of matching impedance to the feed line is made easier [11]. In his February 1935 paper, Brown gave examples of how to calculate the magnitude and phase of earth currents near the base of broadcast antennas and the power dissipation associated with such currents. He went on to discuss the benefits of artificial ground systems of buried conductors and how extensive they needed to be. He also included some experimental data measured using a vertical antenna operated over a copper screen. Brown was born in 1908 in Wisconsin and graduated in electrical engineering from the University of Wisconsin in 1930 [12].

A shortwave antenna is a radio antenna that can be used for transmission (and reception) of any shortwave radio band from among the greater part of the shortwave radio spectrum, without requiring any band-by-band adjustment of the antenna [13]. Generally speaking, there is no difficulty in building an adequate receiving antenna; the challenge is designing an antenna which can be used for transmission without an adjustable impedance matching network [14]. The equations provided below can be used to derive the electric and magnetic field components of a half-wavelength dipole. Many observations have been made by the author on other types of wave projector, in which the antenna and reflectors operate resonantly at a half wavelength. Polar curves thus obtained have proved that a beam with the sharpness never attained could then be produced [15].

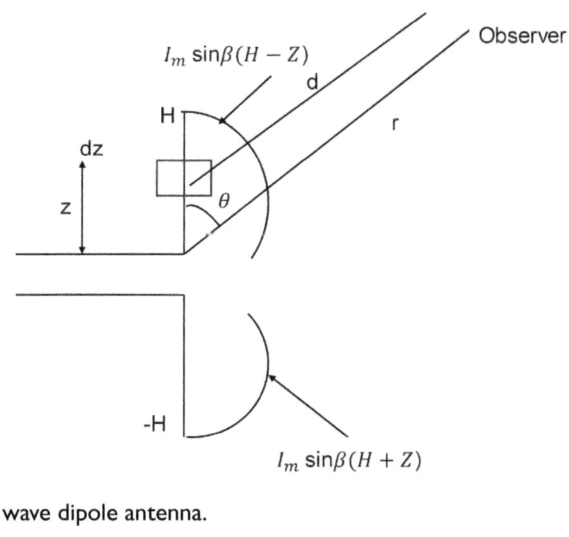

Figure 5.8 Half wave dipole antenna.

Radiated power by a half-wave dipole, $P_T = 73.0\, I_{eff}^2$.

Radiation resistance of half-wave dipole, $R_r = 73\ \Lambda$.
The following procedures are used to get it:

- Develop assertions regarding the element's assumed current distribution.
- Determine the formula for the magnetic potential vector, **A**.
- Derive H from A.
- Obtain **E** from $(\frac{E}{H}) = \eta_0$
- Calculate the mean radiated power P_{av}
- Calculate the overall radiated power.
- Determine the radiation resistance quantity.

Figure 5.2 represents the sinusoidal current distribution.

$$I = I_m \ \sin\beta(H - Z) \quad \text{for } z > 0$$

$$I = I_m \ \sin\beta(H + Z) \quad \text{for } z < 0$$

Here I_m = current maximum.
The potential vector at point P caused by the present element $I\ dz$ is given by

$$dA = dA_z a_z = \frac{\mu_\circ I e^{-j\beta d\, dz}}{4\pi d} a_z \tag{5.40}$$

The distance between the current element and point P is denoted by the variable (d). The total vector potential at point P, expressed as

$$A_z = \frac{\mu_\circ}{4\pi}\int_{-H}^{H} \frac{I\ e^{-j\beta d\ dz}}{d} dz \tag{5.41}$$

$$= \frac{\mu_\circ}{4\pi}\int_{-H}^{0} \frac{I_m \ \sin\beta(h+Z)}{d} e^{-j\beta d} dz + \frac{\mu}{4\pi}\int_{0}^{H} \frac{I_m \ \sin\beta(h-Z)}{d} e^{-j\beta d} dz \tag{5.42}$$

It is crucial to analyze the radiation fields in this context. It is possible to substitute the variable in the denominator with the letter r. During the phase period, the variable d is found in the numerator of the expression.

$$d = r - z\cos\theta$$

Now Equation (5.33) becomes

$$A_z = \frac{\mu_\circ}{4\pi}\int_{-H}^{0} \frac{I_m \sin\beta(h+Z)}{r} e^{-i\beta(r-z\cos\theta)} dz$$

$$+\frac{\mu}{4\pi}\int_{0}^{H} \frac{I_m \sin\beta(h-Z)}{r} e^{-i\beta(r-z\cos\theta)} dz$$

$$= \frac{\mu_\circ I_m e^{-i\beta r}}{4\pi r}\left[\int_{-H}^{0} \sin\beta(h+Z)e^{i\beta z\cos\theta} dz + \int_{0}^{H} \sin\beta(h+Z)e^{i\beta z\cos\theta} dz\right] \tag{5.43}$$

For a half-wave dipole, $h = \frac{\lambda}{4}$

But $\sin\beta(h+Z) = \sin\beta h \cos\beta z + \cos\beta h \sin\beta z$

$\sin\beta(h-Z) = \sin\beta h \ \cos\beta z - \cos\beta h \ \sin\beta z$ As $\beta = \frac{2\pi}{\lambda}$,

$$\sin\beta h = \sin\frac{2\pi}{\lambda}\times\frac{\lambda}{4} = 1,$$

$$\cos\beta h = \cos\frac{\pi}{2} = 0$$

$$\text{So,} \sin\beta(h+Z) = \sin\beta(h-Z) = \cos\beta z \tag{5.44}$$

Equation (5.34) into Equation (5.35), we obtain

$$A_z = \frac{\mu_\circ I_m e^{-i\beta r}}{4\pi r}\left[\int_{-H}^{0} \cos\beta z \ e^{i\beta z\cos\theta} + \int_{0}^{H} \cos\beta z \ e^{i\beta z\cos\theta} dz\right] \tag{5.45}$$

But $\int_{-H}^{0} \cos\beta z \ e^{i\beta z\cos\theta} dz = \int_{0}^{H} \cos\beta z \ e^{i\beta z\cos\theta} dz$

$$A_z = \frac{\mu_\circ I_m e^{-i\beta r}}{4\pi r}\left[\int_{0}^{\lambda/4} \cos\beta z \ (e^{i\beta z\cos\theta} + e^{-i\beta z\cos\theta}) dz\right]$$

$$= \frac{\mu_\circ I_m\, e^{-j\beta r}}{4\pi r}\left[\int_0^{\lambda/4} \cos\{\beta z(1+\cos\theta)\} + \cos\{\beta z(1-\cos\theta)\}\right]dz$$

$$= \frac{\mu_\circ I_m\, e^{-j\beta r}}{4\pi r}\left[\frac{\sin\{\beta z(1+\cos\theta)\}}{\beta(1+\cos\theta)} + \frac{\sin\{\beta z(1-\cos\theta)\}}{\beta(1-\cos\theta)}\right]_0^{\lambda/4}$$

$$= \frac{\mu_\circ I_m\, e^{-j\beta r}}{4\beta\pi r}\left[\frac{(1-\cos\theta)\cos\left(\frac{\pi}{2}\cos\theta\right)+(1+\cos\theta)\cos\left(\frac{\pi}{2}\cos\theta\right)}{\sin^2\theta}\right]$$

$$A_z = \frac{\mu_\circ I_m\, e^{-j\beta r}}{2\beta\pi r}\left[\frac{\cos\frac{\pi}{2}\cos\theta}{\sin^2\theta}\right] \tag{5.46}$$

But, we have

$$\mu_\circ H_\varnothing = \frac{1}{r}\left[\frac{\partial}{\partial r}(r A_\theta) - \frac{\partial}{\partial\theta}A_r\right]$$

$$= \frac{1}{r}\left[\frac{\partial}{\partial r}r(-A_z\sin\theta) - \frac{\partial}{\partial\theta}(A_z\cos\theta)\right]$$

$$\mu_\circ H_\varnothing = -\sin\theta\frac{\partial A_z}{\partial r} \tag{5.47}$$

From Equations (5.38) and (5.39), we have

$$\mu_\circ H_\varnothing = -\frac{\partial}{\partial r}\left(\frac{j\mu_\circ I_m e^{-j\beta r}}{2\beta\pi r}\frac{\cos\left(\frac{\pi}{2}\cos\theta\right)}{\sin^2\theta}\right)\sin\theta$$

$$H_\varnothing = \frac{jI_m e^{-j\beta r}}{2\pi r}\frac{\cos\left(\frac{\pi}{2}\cos\theta\right)}{\sin\theta} \tag{5.48}$$

We also know that $E_\theta = \eta_\circ H_\varnothing$, $\eta_\circ = 120\pi$

$$E_\theta = \frac{j120\pi I_m e^{-i\beta r}}{2\pi r}\left[\frac{\cos\left(\frac{\pi}{2}\cos\theta\right)}{\sin\theta}\right]$$

$$= \frac{j60 I_m e^{-i\beta r}}{r}\left[\frac{\cos\left(\frac{\pi}{2}\cos\theta\right)}{\sin\theta}\right] \tag{5.49}$$

The amplitude of E in the radiation field is

$$E_\theta = \frac{60 I_m}{r}\left[\frac{\cos\left(\frac{\pi}{2}\cos\theta\right)}{\sin\theta}\right] V/m \tag{5.50}$$

E_θ and $H_\varnothing$ are synchronized. Therefore, the greatest value of the Poynting vector is

$$P_{max} = (E_\theta)_{max}\left(H_\varnothing\right)_{max}$$

$$= \frac{60 I_m}{r}\left[\frac{\cos\left(\frac{\pi}{2}\cos\theta\right)}{\sin\theta}\right] \times \frac{I_m}{2\pi r}\left(\left[\frac{\cos\left(\frac{\pi}{2}\cos\theta\right)}{\sin\theta}\right]\right)$$

$$= \frac{30 I_m^2}{\pi r^2}\left[\frac{\cos^2\left(\frac{\pi}{2}\cos\theta\right)}{\sin^2\theta}\right] \tag{5.51}$$

The mean value of the Poynting vector is half of its peak.

$$\text{So}, P_{av} = \frac{15\, I_m^2}{\pi r^2}\left[\frac{\cos^2\left(\frac{\pi}{2}\cos\theta\right)}{\sin^2\theta}\right]$$

$$P_{av} = \frac{\eta_\circ I_m^2}{8\pi^2 r^2}\left[\frac{\cos^2\left(\frac{\pi}{2}\cos\theta\right)}{\sin^2\theta}\right] \tag{5.52}$$

Hence, the total power emitted by a half-wave dipole via a spherical surface is

$$P_T = \oint P_{av} ds = \frac{\eta_\circ I_m^2}{8\pi^2 r^2}\int_0^\pi \frac{\cos^2\left(\frac{\pi}{2}\cos\theta\right)}{\sin^2\theta} 2\pi r^2 \sin\theta d\theta$$

$$= \frac{\eta_\circ I_m^2}{4\pi}\int_0^\pi \frac{\cos^2\left(\frac{\pi}{2}\cos\theta\right)}{\sin\theta} d\theta \tag{5.53}$$

But the numerical evaluation of the integral $\int_0^\pi \frac{\cos^2\left(\frac{\pi}{2}\cos\theta\right)}{\sin\theta} d\theta$ Using Simpson's or the trapezoidal rule yields a value of 1.218.

So,

$$P_T = \frac{\eta_\circ I_m^2}{4\pi}$$

$$= \frac{120\pi\, I_m^2}{4\pi} \times 1.218 = 36.54 I_m^2 \tag{5.54}$$

As $I_m = \sqrt{2} I_{eff}$, Equation (5.55) becomes

$$P_T = 36.54 \times 2 \times I_{eff}^2$$

$$P_T = 73.08\, I_{eff}^2 \text{watts} \tag{5.55}$$

The coefficient of I_{eff}^2 is the radiation resistance, specifically,

$$R_r = 73.08 \tag{5.56}$$

5.9 ANALYSIS OF RADIATION PATTERN OF A HALF-WAVE DIPOLE IN MATLAB

It has been determined with the use of MATLAB that the radiation pattern of a half-wave dipole antenna has been examined.

MATLAB Code is given below:

```
clear all
for ii= 1:360
  radian(ii)= ii*pi/180;
  mag(ii)= abs((cos(pi/2*cosd(ii)))/(sind(ii)));
  edb(ii)= 20*log 10 (mag(ii));
  if edb(ii) < -100  % to avoid infinity
    edb(ii) = -100;
  end
end
deg = radian* 180/pi;
subplot (1,2,1);
plot (deg,edb,'r');
subplot (1,2,2);
% Define theta values from 0 to 180 degrees
theta = linspace(0, pi, 1000);

% Calculate the radiation pattern for a half-wave dipole antenna
E_theta = abs(sin(theta));

% Plot the radiation pattern in polar coordinates
polarplot(theta, E_theta, '-b');

% Add title and labels
title ('Radiation Pattern of Half-Wave Dipole Antenna');
```

The simulation results of the half-wave dipole in terms of a rectangular plot are shown in Figure 5.9 while the polar plot is shown in Figure 5.10.

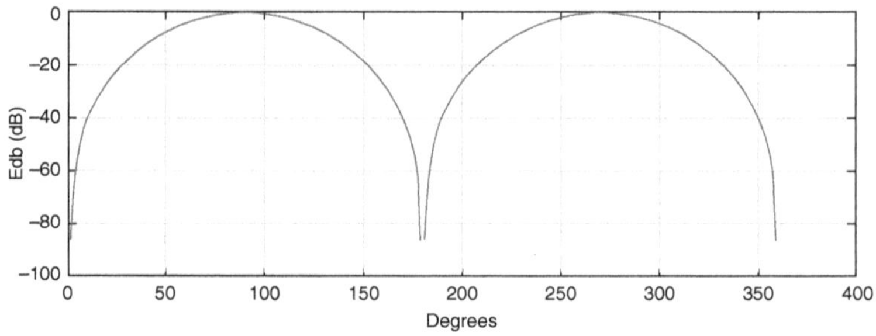

Figure 5.9 Rectangular plot of half-wave dipole.

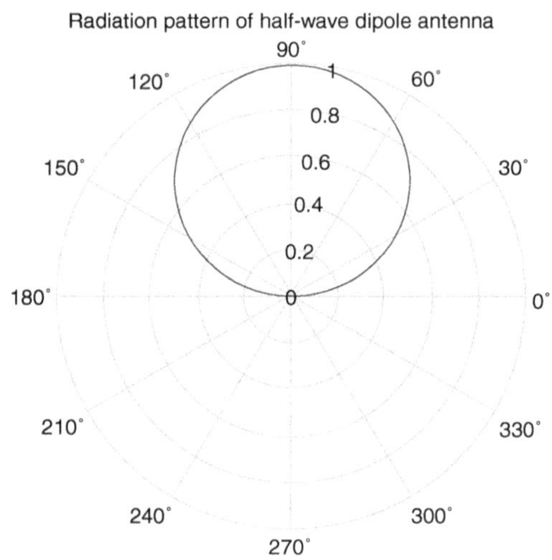

Figure 5.10 Polar plot of half-wave dipole.

5.10 RADIATION PATTERN FROM QUARTER-WAVE MONOPOLE

The power output of the quarter-wave monopole, $P_T = 36.5\, I_{eff}^2$ watts

Radiation resistance, $R_r = 36.5\Omega$.

Examine Figure 5.2, which depicts a monopoly with a current distribution.

Acquire *P-a**X*** precisely as mentioned in the half-wave dipole. In other words, as shown in Equation (5.44), we refer to Figure 5.2, for illustrating a monopoly with its existing distribution.

$$P_{av} = \frac{\eta_\circ I_m^2}{8\pi^2 r^2}\left[\frac{\cos^2\left(\frac{\pi}{2}\cos\theta\right)}{\sin^2\theta}\right] \tag{5.57}$$

The monopoly only spreads over a hemispherical surface due to being connected to a perfectly conducting plane at one end. Therefore, the overall power emitted is

$$P_T = \oint P_{av} ds$$

$$= \frac{\eta_\circ I_m^2}{8\pi r^2}\int_0^{\frac{\pi}{2}} \frac{\cos^2\left(\frac{\pi}{2}\cos\theta\right)}{\sin^2\theta} 2\pi r^2 \sin\theta d\theta$$

$$= \frac{\eta_\circ I_m^2}{4\pi}\int_0^{\frac{\pi}{2}} \frac{\cos^2\left(\frac{\pi}{2}\cos\theta\right)}{\sin^2\theta} d\theta$$

Numerical evaluation of the integral $\int_0^{\frac{\pi}{2}} \frac{\cos^2\left(\frac{\pi}{2}\cos\theta\right)}{\sin^2\theta} d\theta$ by Simpson's or trapezoidal rule gives a value of 0.609.

So $P_T = \frac{\eta_\circ I_m^2}{4\pi} \times 0.609 = 18.27 I_m^2$

As $I_m = \sqrt{2} I_{eff}$

$$P_T = 36.54 \times 2 \times I_{eff}^2 \tag{5.58}$$

The radiation resistance

$$R_r = 36.54 \tag{5.59}$$

5.11 CONCLUSION

This chapter provides a description of microstrip antennas and their parametric analysis. Additionally, the chapter presents the results of the gain and radiation pattern. Microstrip antennas offer superior performance in integrated, compact, and planar applications with regulated radiation patterns. Dipole antennas, on the other hand, are characterized by their simplicity, ease of use, and omnidirectional radiation, which is particularly useful for providing extensive coverage. In addition, microstrip antennas can modify radiation patterns. Dipole antennas, with their radiation pattern, S11 parameter, and radiation mechanisms, provide a unique set of advantages tailored to specific application requirements. When engineers have a better understanding of these antennas' dynamics and radiation mechanisms, they are better able to modify their designs to meet specific requirements. This enhances the performance of contemporary communication systems. When researchers gain a better understanding of the dynamics and radiation mechanisms of these antennas, they can modify their designs to meet specific requirements. This is because they are more capable of adjusting their designs. As a result, contemporary communication networks' efficiency demonstrates significant improvement.

REFERENCES

[1] "WHO Ageing and Health Fact Sheet," 2022. [Online]. Available: www.who.int/news-room/fact-sheets/detail/ageing-and-health

[2] V S. Duti Rekha, R. Gude, N. B. Bobinath, A. Kumar, S. Samanth, and S. Majji, "Healthcare Monitoring Wearable Microstrip Patch Antenna for Elder People with Acceptable SAR," *2023 Second International Conference on Electrical, Electronics, Information and Communication Technologies (ICEEICT), Trichirappalli, India*, IEEE, 2023, pp. 1–4, doi: 10.1109/ICEEICT56924.2023.10157472.

[3] K. Zhang, P. J. Soh, and S. Yan, "Design of a Compact Dual-Band Textile Antenna Based on Metasurface," *IEEE Transactions on Biomedical Circuits and Systems*, vol. 16, no. 2, pp. 211–221, 2022, doi: 10.1109/TBCAS.2022.3151243.

[4] Y. Chen, X. Liu, Y. Fan, and H. Yang, "Wearable Wideband Circularly Polarized Array Antenna for Off-Body Applications," *IEEE Antennas and Wireless Propagation Letter*, vol. 21, no. 5, pp. 1051–1055, 2022, doi: 10.1109/LAWP.2022.3157367.

[5] W. L. Stutzman and G. A. Thiele, *Antenna Theory and Design*, 2nd ed., New York: Wiley, 1998, p. 172, 173.

[6] E. C. Jordan and K. G. Balmain, *Electromagnetic Waves and Radiating Systems*, New Delhi: Prentice-Hall of India Pvt. Ltd., 1976.

[7] C. A. Balanis, "Antenna Theory: A Review," *Proceeding of the IEEE*, vol. 80, no. 1, pp. 7–23, 1992, doi: 10.1109/5.119564.

[8] O. P. Gandhi, *Microwave Engineering and Applications*, New York: Pergamon, 1985.
[9] M. F. Iskander, *Electromagnetic Fields and Waves*, Englewoof Cliffs, NJ: Prentice-Hall, 1992.
[10] J. D. Kraus, *Antennae*, 2nd ed., New York: McGraw-Hill, 1988.
[11] D. M. Pozar, *Microwave and RF Design of Wireless Systems*, John Wiley and Sons Inc., 2001, p. 134. ISBN 978-0-471-32282-5.
[12] H. Yagi, "Scanning the Past: A History of Electrical Engineering from the Past," IEEE. *Reprinted with permission from the IEEE publication*, vol. 81, no. 6, 1993.
[13] R. C. Johnson and H. Jasik (eds.), *Antenna Engineering Handbook*, New York: McGraw-Hill, 1984.
[14] U. Shozo and S. Gentei, "Directive Short Wave Antenna, 1924," *IEEE Milestones*, IEEE History Center, IEEE, 2005.
[15] S. Uda, "High Angle Radiation of Short Electric Waves," Proceedings of the IRE, vol. 15, pp. 377–385, May 1927.

Chapter 6

28/38 GHz millimeter wave MIMO antenna for 5G wireless communication systems

F. B. Shiddanagouda and P. S. Aithal

6.1 INTRODUCTION

In today's wireless landscape, there is a surging demand for compact wireless devices that can handle emerging technologies such as artificial intelligence (AI), data science (DS), cyber security (CS), and the internet of things (IoT), while also delivering high-speed gigabit data rate transmission [1–3]. Addressing these demands has led to the exploration of various systematic solutions across different layers of the Transmission Control Protocol/Internet Protocol (TCP/IP) modules, as documented in the literature. Among NG these solutions, millimeter wave Multiple Input Multiple Output (MIMO) antenna systems emerge as promising candidates. A critical challenge in millimeter wave MIMO antenna systems is to effectively manage multiple antennas within the same ground plane. This often results in significant mutual coupling issues between the antenna elements, leading to a degradation in antenna performance. To mitigate these challenges and enhance antenna parameters, researchers have explored various techniques documented in the literature [4–5]. These techniques include photonic band gap structures, metamaterial structures, and defected ground structures (DGSs). This particular study focuses on the design of a four-element millimeter wave MIMO antenna utilizing diagonal square slot DGSs. The diagonal square slot DGS serves as a technique to overcome mutual coupling issues and improve antenna performance. In the forthcoming sections, the paper systematically explores the antenna layout design framework, presents results, and draws conclusions.

6.2 APPROACHES TO DESIGN MIMO ANTENNA

Initially, a resonating frequency of 38GHz was selected for the single-millimeter wave antenna (SMA). The antenna was designed using RT/duroid 5880LZ substrate material, and its specifications are shown in Table 6.1. The SMA structures were designed utilizing an HFSS EM simulator. They

DOI: 10.1201/9781003470854-6

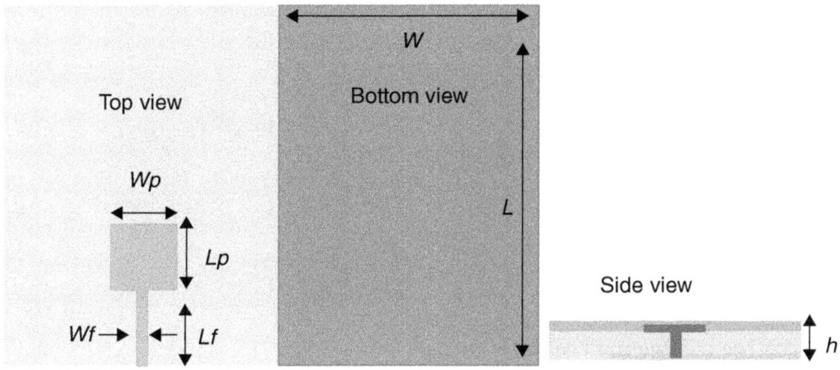

Figure 6.1 The configuration of SMA.

Table 6.1 Specifications of the substrate material

Material	RT/Duroid 5880LZ
Relative Permittivity	2.2
Loss Factor	0.0009
Thickness	0.8mm

Table 6.2 Dimensions of SMA

Variables	*L*	*W*	*Lp*	*Wp*
Dimensions	10mm	10mm	2.02mm	3.12mm
Variables	Lf	Wf		h
Dimensions	2.1mm	0.4mm		0.8mm

were fed with a 50Ω microstrip line, as depicted in Figure 6.1. Detailed antenna dimensions are provided in Table 6.2 [6].

The SMA has been expanded to a four-element millimeter wave antenna in MIMO configuration. This configuration consists of four identical millimeter wave antennas with a λ/4 distance maintained to prevent coupling effects between the antenna elements. Enhancing efficiency ground plane size set to 20 mm × 20 mm. The configuration of the four-element millimeter wave MIMO Antenna (MMA) is presented in Figure 6.2.

To enhance the performance of the aforementioned millimeter wave MIMO antenna (MMA), further investigation was conducted by employing a technique known as DGS. This method involves etching the ground plane, aiming to optimize antenna performance [7]. During the

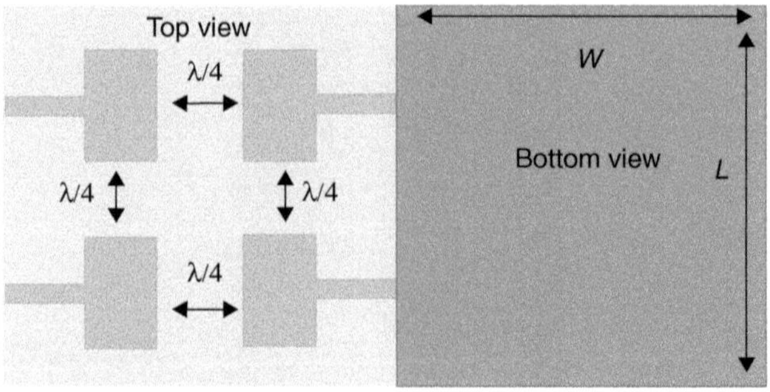

Figure 6.2 The configuration of MMA.

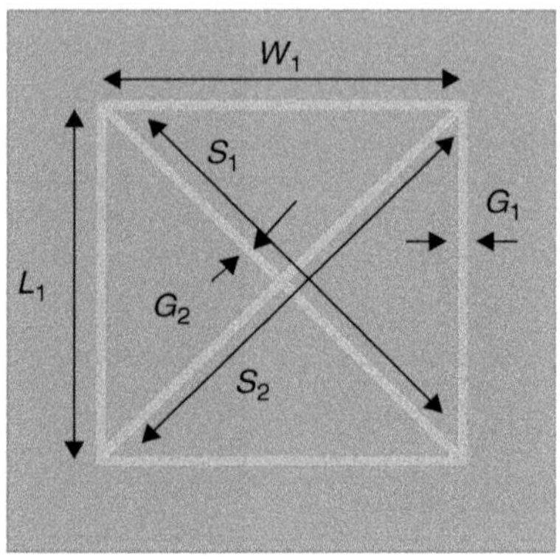

Figure 6.3 The configuration of DSSDGS.

Table 6.3 Dimensions of DSSDGS

Variables	*Dimensions*
W_1	3.8 mm
L_1	3.4 mm
S_1, S_2	5.37 mm
G_1,G_2	0.15 mm

design phase, a millimeter-wave MIMO antenna was meticulously crafted by employing a diagonal square slot defected ground structure (DSSDGS). The DSSDGS unit cells, detailed in Figure 6.3 and optimized in Table 6.3, were integrated into the antenna's ground plane. By etching five DSSDGS unit cells with 2.75mm spacing, intentional disruption of the current distribution in the ground plane was achieved. This deliberate intervention

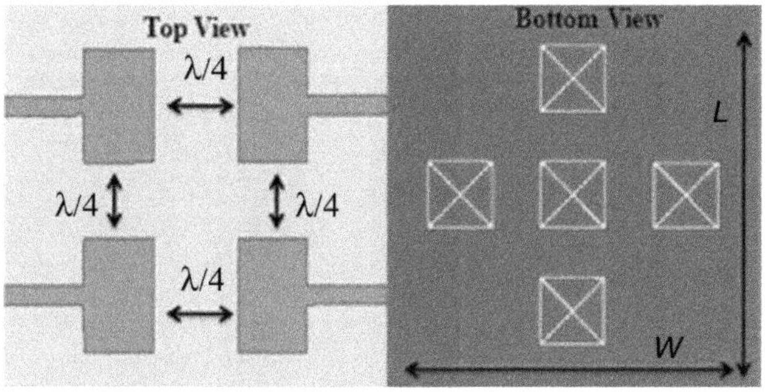

Figure 6.4 The configuration of MMMA.

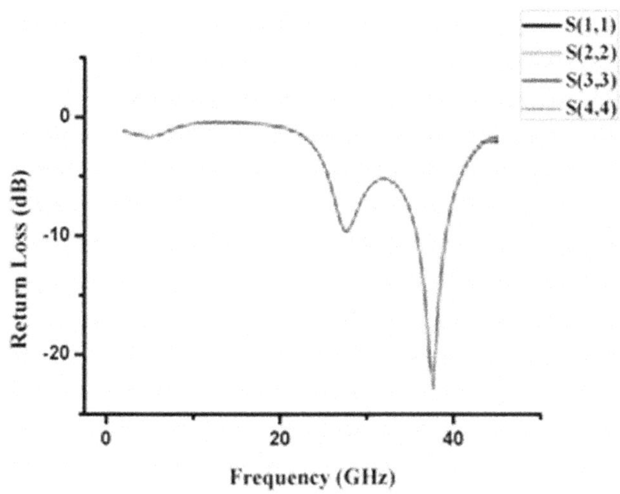

Figure 6.5 The return loss characteristics associated with MMA.

resulted in size reduction, dual-band resonance, and significant decreases in mutual coupling within the antenna system. The modified millimeter-wave MIMO antenna (MMMA) structure, shown in Figure 6.4, embodies these enhancements.

6.3 EXPLORING MIMO ANTENNA OUTCOMES

The carefully scrutinized return loss properties of the millimeter wave MIMO antenna (MMA) are vividly depicted in Figure 6.5. Notably, the MMA sharply resonates at 38GHz, showcasing an exceptional return loss of –22.18dB, along with an impressive bandwidth spanning over 3GHz.

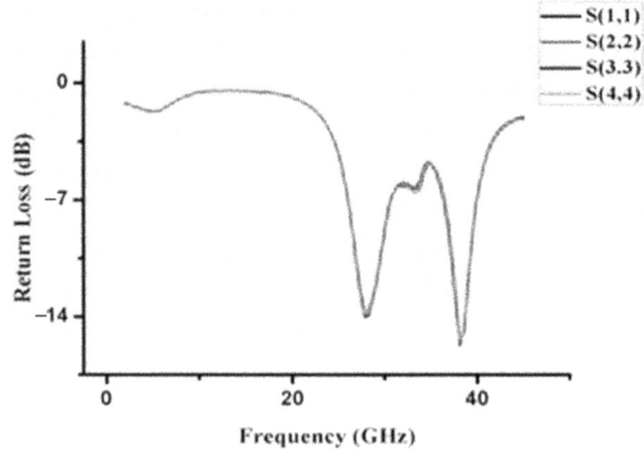

Figure 6.6 The return loss characteristics associated with MMMA.

The MMMA exhibits several return loss characteristics, as depicted in Figure 6.6. By incorporating five DSSDGS structures with a 2.75mm separation distance on the ground plane, the antenna effectively suppresses unwanted surface waves. Consequently, the antenna resonates at dual frequencies, 28GHz and 38GHz, with return loss values of −13.95dB and −15.96dB, respectively. At these resonance frequencies, the antenna provides bandwidths of 3.8GHz and 3.5GHz, respectively. Moreover, this modification achieves maximum size reductions, which can be quantified using the following equation:

$$\text{Size reduction } (\%) = \left(\frac{\text{Length}_1 - \text{Length}_2}{\text{Length}_1} \right) \times 100 \tag{6.1}$$

The MMMA achieved a 32.33% reduction in size by utilizing five DSSDGS unit cells. This size reduction is quantified by comparing the lengths of the smaller resonating frequency (28GHz) denoted as Length1, and the length of the larger resonating frequency (38GHz) denoted as Length2.

When multiple antennas are situated on a shared ground plane, they can influence each other due to electromagnetic interactions, known as mutual coupling effects. These effects can lead to changes in the antenna's performance, such as impedance mismatch or radiation pattern distortion. The mutual coupling coefficients (MCC) quantify the strength of these interactions between different antenna elements. By utilizing the given equation, we can compute these coefficients, providing a valuable understanding of the level of coupling and how it influences the antenna system's overall performance [8].

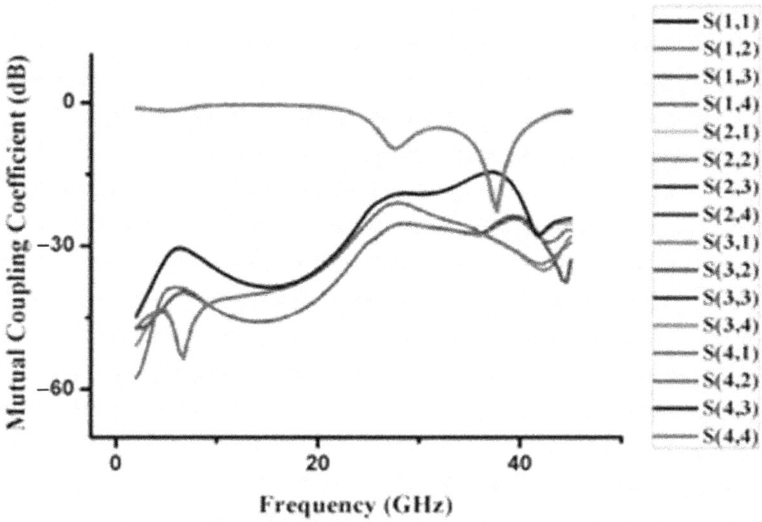

Figure 6.7 MCC of MMA.

$$S_{12} = 20\log\left(\frac{2Z_{21} - Z_0}{[(Z_{11} + Z_0)^2 + (Z_0)^2]}\right) \tag{6.2}$$

In Figure 6.7, the MCC characteristics of anMMA are illustrated, highlighting an MCC of –14.64dB at 28GHz. This figure demonstrates a strong coupling between return loss parameters (*S* parameters) such as S_{11}, S_{22}, S_{33}, S_{44}, and mutual coupling terms (S_{12}, S_{13}, S_{14}, S_{21}, S_{23}, S_{24}, S_{31}, S_{32}, S_{34}, S_{41}, S_{42}, S_{43}) are examined at the resonating frequency of 28GHz, which needs to be minimized for optimal antenna performance.

Figure 6.8 illustrates the effect of employing five DSSDGS structures on the ground plane, showing no overlap between the *S*-parameters and mutual coupling terms at their respective resonating frequencies of 28GHz and 38GHz. This absence of overlap leads to a notable decrease in MCC.

As a result, the MMMA depicted in Figure 6.8 demonstrates improved performance with MCC values of –18.67dB and –23.12dB at 28GHz and 38GHz, respectively. This reduction in mutual coupling enhances the antenna's efficiency and signal isolation, contributing to overall better performance in MIMO communication systems.

The total peak gain (TPG) of an antenna is crucial for calculating the antenna link budget and coverage area. The MMA achieves a TPG of 4.85dB at 28GHz, as depicted in Figure 6.9. In comparison, the MMMA

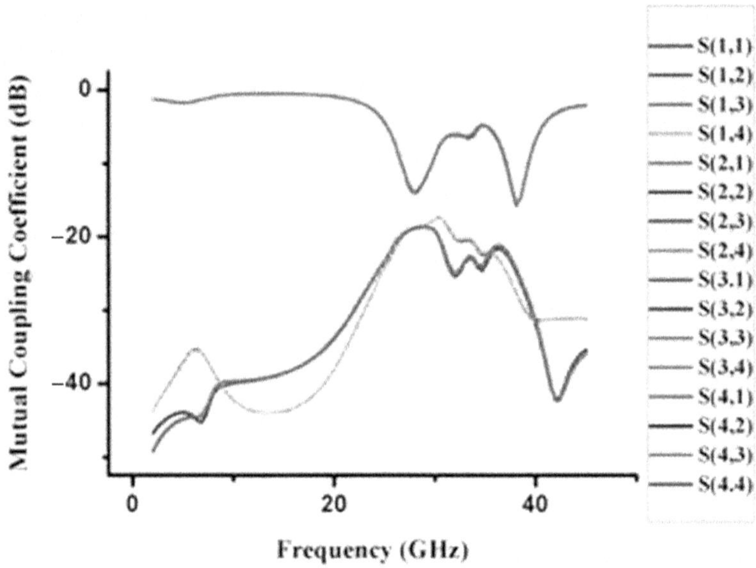

Figure 6.8 MCC of MMMA.

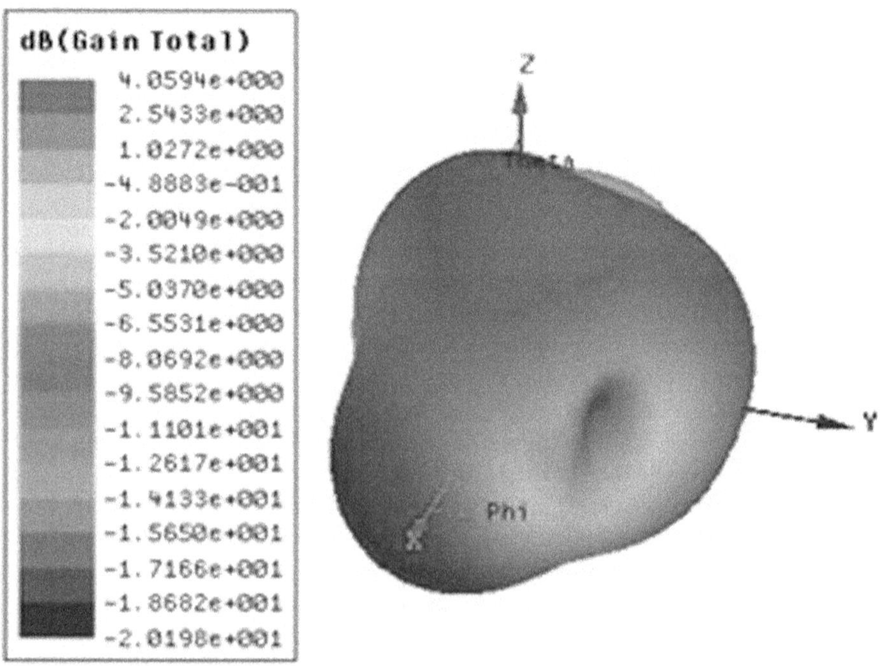

Figure 6.9 TPG of MMA.

demonstrates higher TPG values of 5.32dB at 28GHz and 5.45dB at 38GHz, shown in Figure 6.10. These improved TPG values suggest that the MMMA can better concentrate radiated power, potentially leading to enhanced performance in terms of signal propagation and coverage area.

The radiation pattern demonstrates the propagation of electromagnetic waves toward the far field region. Figure 6.11illustrates the radiation characteristics of the millimeter wave MIMO antenna, particularly focusing on its broadside radiation pattern. On the other hand, Figure 6.12 presents the radiation characteristics of the modified millimeter wave MIMO antenna. Similar to the original antenna, it also exhibits broadside radiation at both resonating frequency points. This implies that the modified antenna maintains its ability to radiate energy efficiently perpendicular to its surface, suggesting consistent performance even after modifications.

Envelope correlation coefficients (ECC)are used to assess the isolation of communication channels [9]. In Equation (6.3), the millimeter-wave MIMO antennaexhibits an ECC of 0.000001, as depicted in Figure 6.13. However, the modified millimeter-wave MIMO antenna achieves an ECC of 0 at the respective frequency points, as illustrated in Figure 6.14. This suggests that the modified antenna effectively minimizes channel correlation, leading to improved isolation between communication channels.

$$\rho = \frac{\left|S_{11}^{*}S_{12} + S_{21}^{*}S_{22}\right|^{2}}{\left(1-\left|S_{11}\right|^{2}-\left|S_{21}\right|^{2}\right)\left(1-\left|S_{22}\right|^{2}-\left|S_{12}\right|^{2}\right)} \tag{6.3}$$

Diversity gain (DG), a crucial metric for assessing MIMO antenna efficiency, relies on ECC for its calculation [10–12]. This relationship underscores the significance of ECC in determining the effectiveness of MIMO antennas. By evaluating ECC, the DGprovides valuable insights into the antenna's ability to enhance signal diversity and mitigate channel correlation effects, thereby improving the overall performance of the MIMO system.

$$DG = 10\sqrt{\left(1-\left|\rho\right|^{2}\right)} \tag{6.4}$$

The DG of a millimeter-wave MIMO antenna measures its ability to improve signal diversity and reduce channel correlation. In this case, the millimeter-wave MIMO antenna achieves a DG of 9.99dB, while the modified version reaches a DG of 10dB at both frequency points. This indicates that the modification enhances signal diversity even further, resulting in

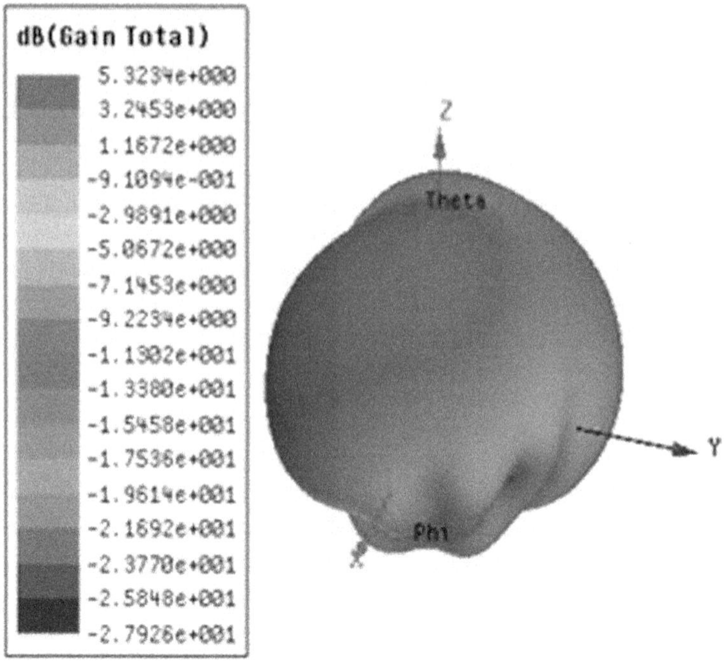

(a) 5.32 dB at 28 GHz

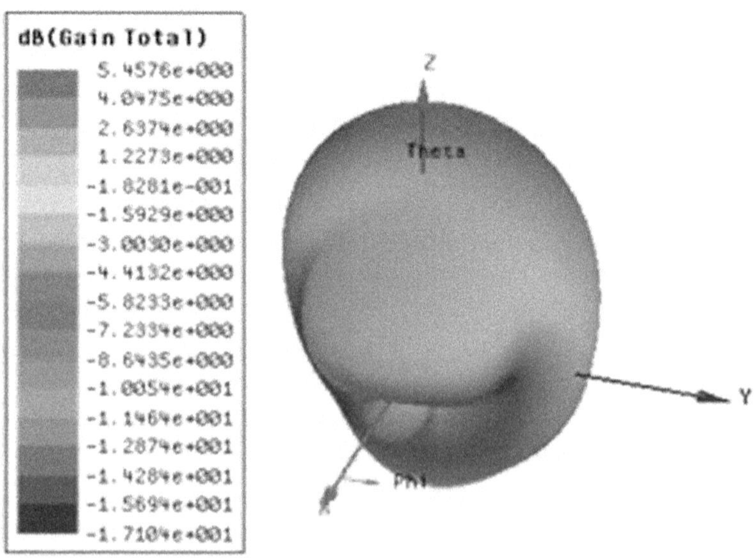

(b) 5.45 dB at 38 GHz

Figure 6.10 (a) and (b) present the TPG of MMMA.

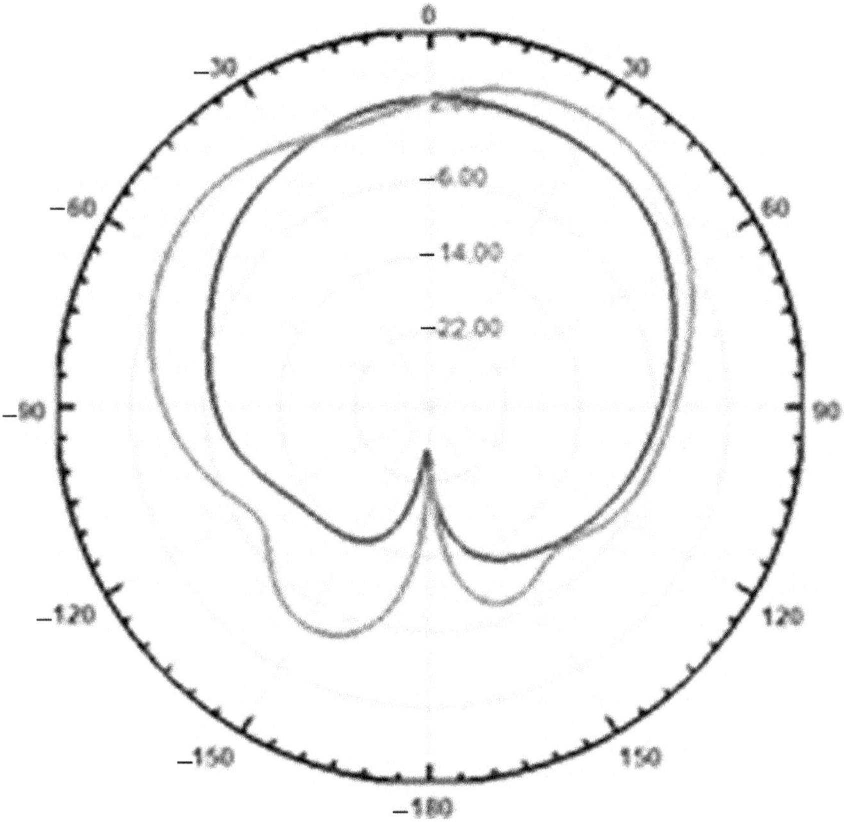

Figure 6.11 Radiation pattern of MMA.

slightly improved performance compared to the original antenna. The results for both antennas are summarized in Table 6.4.

As shown in Table 6.4, a modified millimeter wave MIMO antenna utilizing a DSSDGS with five-unit cells resonates at dual frequency bands of 28GHz and 38GHz. The antenna boasts a bandwidth of 3.8GHz and 3.5GHz, withMCC of –18.67dB and –23.12dB at their respective frequency points. Additionally, this antenna offers a virtual size reduction of 32.33% and a DG of 10dB.These results indicate that the MMMA is well-suited for 5G wireless communication systems.

6.4 CONCLUSIONS

The antennas were developed using the HFSS electromagnetic simulator, leveraging a diagonal square slot defected ground structure (DSSDGS)

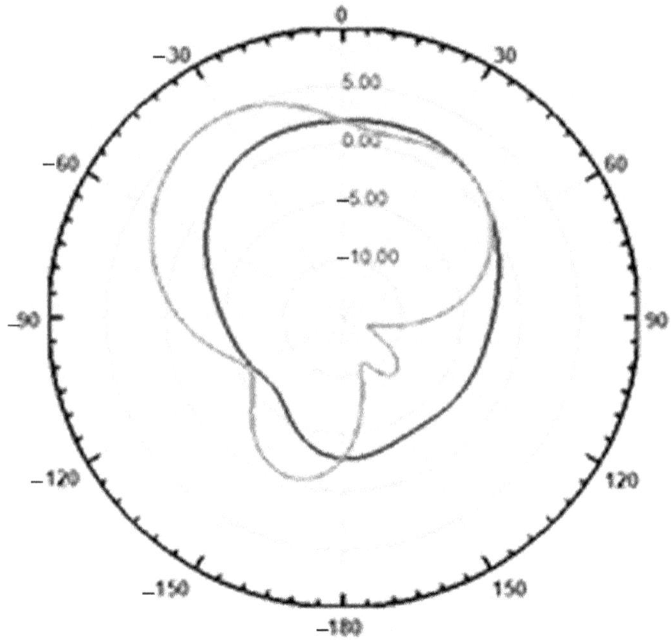

(a) 28 GHz

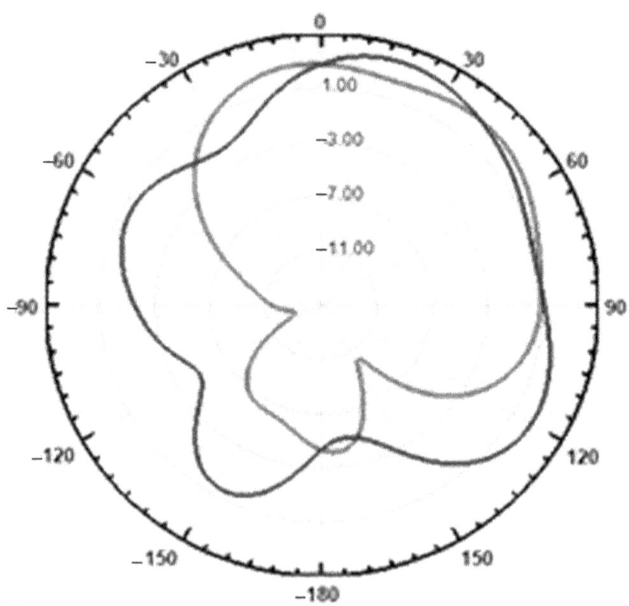

(b) 38 GHz

Figure 6.12 (a) and (b) showcase the radiation patterns for MMMA.

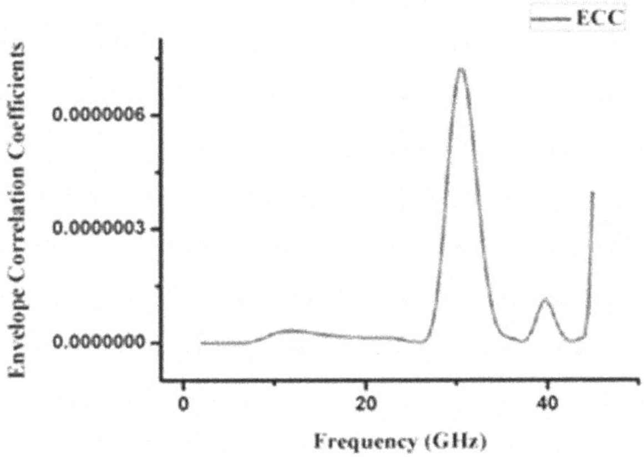

Figure 6.13 ECC of MMA.

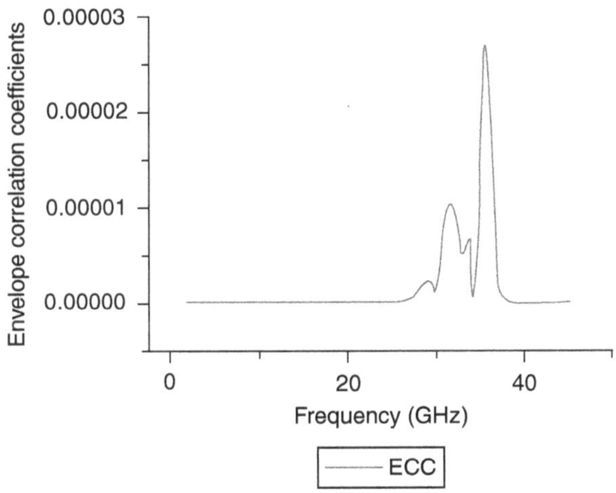

Figure 6.14 ECC of MMMA.

to enhance MIMO antenna performance. The modified millimeter-wave MIMO antenna (MMMA) resonates at 28GHz and 38GHz with return loss values of –13.95dB and –15.96dB, respectively. It boasts a wide bandwidth of 3.8GHz and 3.5GHz, along with impressive MCC of –18.67dB and –23.12dB. Furthermore, notable progress was achieved in TPG, envelope correlation coefficients, and DG at both frequency points. Additionally, a significant 32.33% reduction in virtual size was observed. These outcomes

Table 6.4 Tabulated results of MMA and MMMA

Specifications	*MMA*	*MMMA*	
Resonating Frequency (GHz)	38	28	38
Return Loss (dB)	−22.18	−13.95	−15.76
Bandwidth (GHz)	3	3.8	3.5
MCC (dB)	−14.64	−18.67	−23.12
TPG(Total Peak Gain) (dB)	4.85	5.32	5.45
Size Reduction (%)	–	32.33%	
ECC (Envelope Correlation Coefficient)	0.000001	0	
DG (Diversity Gain) (dB)	9.9	10	

underscore the appropriateness of the developed millimeter-wave MIMO antenna for 5G wireless communication systems, showcasing its potential for high-performance applications.

REFERENCES

1. R. Patel and A. Desai, "An electrically small antenna using defected ground structure for RFID, GPS, and IEEE802.11 a/b/g/s applications," *Journal of Progress in Electromagnetic Research Letters*, vol.75, pp.75–81, 2018.
2. Z. Pi and F. Khan, "An introduction to millimeter-wave mobile broadband systems," *IEEE Communications Magazine*, vol. 49, no. 6, pp. 101–107, June 2011, doi: 10.1109/MCOM.2011.5783993.
3. R. Zahn and P. Shutie, "Advanced antenna technologies for X band SAR,"*IEEE Transactions on Antenna and Propagation*, vol, AP-95, no.2, 1995.
4. T. S. Rappaport et al., "Millimeter wave mobile communications for 5G cellular: It will work!," *IEEE Access*, vol. 1, pp. 335–349, 2013.doi: 10.1109/ACCESS.2013.2260813.
5. P. Adhikari, "Understanding millimeter wave wireless communication," Loea Corporation, 2008.
6. Balanis, "Theory of antennas," *IEEE Transactions on Antenna and Propagation,* vol. AP-41, no.9, 1993.
7. F. B. Shiddanagouda, R. M. Vani,and P. V. Hunagund, "Miniaturized dual-band printed MIMO antenna with DGS for L-band and C-band wireless applications," 2020 IEEE International Conference on Communication, Networks and Satellite (Comnetsat), 2020, pp. 111–116, IEEE, doi: 10.1109/Comnetsat50391.2020.9328996.
8. P. Jeyakumar, et al., "Design and simulation of directive high gain microstrip array antenna for 5G cellular communication," Asian Journal of Applied Science and Technology, vol. 2, no. 2, pp. 301–313, 2018.
9. Xiang Cheng and Boyu, "Communication in the real world 3D MIMO,"*IEEE Wireless Communication*, vol. AP-14, no.25, 2014.

10. W. A. Awan, et al., "Patch antenna with improved performance using DGS for 28GHz applications," 2019 International Conference on Wireless Technologies, Embedded and Intelligent Systems (WITS), Fez, Morocco, 2019, pp.1–4, IEEE.
11. M. Kumar and V. Nath, "Analysis of low mutual coupling compact multi band compact antenna and its array using defected ground structure," *Engineering Science and Technology an International Journal*, vol. 19, no. 2, pp. 866–874, 2016.
12. S. Prasad, et al., "Design of CSRR loaded multiband slotted rectangular patch antenna," 2017 IEEE Applied Electromagnetic Conference (AEMC), Aurangabad, 2017, pp.1–2, IEEE.

Chapter 7

Pattern synthesis of a linear antenna array using sooty tern optimization algorithm

Anoop Raghuvanshi, Abhinav Sharma, and Abhishek Kumar Awasthi

7.1 INTRODUCTION

An arrangement of radiating elements arranged in a geometric or electrical configuration makes up an antenna array. In most of the cases, the elements used in making the antenna array are the same in nature. Various geometrical configurations have been used to make an array; here we see a linear antenna array (LAA), which is the special case of a planar array system. Directive and steerable antenna array systems are in high demand for current wireless communication system trends. Considering this maximum energy can be transmitted and received in certain directions while being suppressed in other directions by managing the side lobe level (SLL).

In an antenna array, the radiation pattern is computed by multiplying the radiation pattern of single elements by the array factor. The element's amplitude and phase excitations of every element as well as the distance between them may be tuned for the sake of generating the appropriate radiation characteristics. Pattern synthesis is the process of estimating an antenna array's parameters, such as spacing between the elements, excitation current amplitude and its phase, to produce the required radiation pattern. The radiation characteristics of the array should have fewer side lobes, a shorter beam width, and deep nulls placed in the direction of interference signals.

In the past few decades, many evolutionary algorithms have been applied for array optimization and its synthesis. This technique reduces the search time for finding the exact weights to attain a better array design. To get the desired response, change the ratios of the amplitude, phase, or both between components of an array. Null placement and SLL minimization in LAA are accomplished by the optimization of amplitude and phase of excitation current with the constraint that spacing between the elements is fixed or by the optimization of separation between the elements considering fixed excitation current amplitude and phase. For the synthesis of LAA various metaheuristic algorithms have been tried such as particle swarm optimization (PSO), genetic algorithm (GA), flower pollination algorithm (FPA), moth flame optimization (MFO), grey wolf optimization (GWO),

DOI: 10.1201/9781003470854-7

cuckoo search (CS), cat swarm optimization (CSO), central force optimization (CFO), chicken swarm optimization, ant colony optimization (ACO), ant lion optimization (ALO), and many more. Still, there is a high demand for applying more algorithms to achieve better results.

The side lobes of the radiation pattern are minimized to receive the data in the desired direction, and the half-power beam width is shortened to communicate with low loss across large distances if possible as electromagnetic interference is prevented in the environment [1]. The construction of a non-uniform stimulated symmetric LAA with the optimized spacings of non-uniform in nature, between the components is discussed in the study utilizing Real Coded Genetic Algorithm (RGA) optimization methods [2]. Additionally, as the antenna array's element count grows, so does the amount of time the algorithm takes to compute the amplitude and phase values of each element. The element amplitudes are changed via adjustable attenuators used in the amplitude-alone control. The attenuator count and the time of computation are both reduced by half if there is an even symmetry in the element's amplitudes around the center of the array. When only position is controlled, it requires a mechanical drive mechanism like servomotors for moving the elements of the array, the phase-only control makes use of phase shifters [3]. While studying foraging behavior in humpback whales, it is observed that they search the small fishes in their surrounding area by generating a special kind of bubbles along a circle. This is used in the development of a whale optimization algorithm in reference [4]. The target fish is treated as the best solution for the current. In the initial situation, the best entity is decided and after that, the positions of the rest are upgraded toward the best one.

Cost, fitness, or objective function can be called an efficiency computing parameter which actually tells the closeness of the obtained and actual standard solution for any problem. In this problem of antenna array pattern synthesis, this function guides the obtained solution toward the desired or required antenna radiation pattern [5]. The modified PSO called comprehensive learning particle swarm optimizer (CLPSO) is used, which offers a bigger prospect than the original PSO to investigate more of the search space. The divergence is enhanced with the increment in individual particles searching space leading to better searching the space for computing the global optimum value [6].

In pattern synthesis of an antenna array, the major concern is to estimate the values of current amplitudes near the desired one which makes the desired radiation pattern. This task of searching for the optimum and achieving a low value of SLL will be treated as a problem of non-linear optimization. To tackle such kinds of problems metaheuristic methods are the most used which do not stuck in local optima [7]. The mutual coupling prevailed in an antenna array affects its performance by altering the terminal impedances and the reflection coefficients. The border elements of the linear and rectangular antenna array lead to mutual coupling which is less observed in circular-shaped arrays [8].

CSO is imposed for estimating the optimum position of the antenna elements for the suppression of side lobes and to attain the prescribed nulls. This algorithm is based on the behavioral aspect of the cat. The two phases of the cat that are utilized in the development of this algorithm are seeking and tracing. Mostly the cats are found in rest mode, but they are in full alert mode if they encounter any disturbance near them. They save energy in the rest mode and use that energy while picking their targets [9]. The initial step requires choosing an appropriate array and creating an appropriate fitness function, which is the primary condition in the optimization process [10]. For the improvement of radiation performance of the LAA MFO is applied in reference [11]. The moth belongs to the butterfly family and there is a tendency for them to they attracted toward the light. Based on this natural phenomenon MFO is proposed.

The cross-entropy (CE) method is applied for the designing of a planar array [12] while the composite differential evolution algorithm (CoDE) is applied in reference [13] for the optimization of the element spacing in LAA. Efficient performance is achieved in the context of power utilization and spectral efficiency by maximizing the gain which in turn improves the received signal-to-noise ratio [14]. Constrained and unconstrained optimization problems are solved by applying the ALO which is primarily depending upon the hunting mechanism of the antlions [15]. GWO as applied in reference [16] motivated by the leadership quality and hunting phenomena of the grey wolf. The runner root algorithm (RRA) is motivated by the intelligence of the plant as the plants find the better geographical area for their sustainability. There is a natural mechanism of reinitialization in this algorithm which finds the best means to come out of local minima stagnation and hence the two phases explore, and exploit works independently [17].

The pattern search method is applied in reference [18] in which the algorithm briefly highlights that it estimates the solution that comes closer to the desired one after each iteration. The method looks for a point in the mesh—a collection of points—around the current position at each step until it locates a point where the fitness function value drops relative to the value at the current point. In the following algorithmic step, this new point is set as the current point. A scalar multiple of a predetermined set of vectors known as the "Pattern Vector" is added to the current location to create the mesh.

In an antenna array system whenever any of the antenna elements fail, the field characteristics of the system are severely compromised. When the direction of nulls needs to be preserved in the array under consideration, the complexity of the issue is further raised. The typical bat algorithm's performance is improved in the work proposed in reference [19] using the inertia weight technique. By optimizing the amplitude of the remaining non-defective elements to regain the antenna array's original pattern, an improved bat method is suggested and used to address a real-world problem with a LAA. The rate of success in the bat acts as a feedback parameter for evaluating the state of the bat in the search arena; the adaptive inertia weight scheme is based on this feedback variable. The adaptive inertia weight parameter is then changed as and when required.

Different strategies have been used successfully to optimize antenna arrays [20]. There are numerous local minima in the non-linear array optimization problem. The method of gradient is not a suitable option for these issues because they significantly rely on the starting guess and may stuck in the local minima if the guess is poor. Due to this, stochastic global algorithms are now used to find the global minima while escaping local minima.

A group of bees can be viewed as a distributed organism that can spread out and use several food sources over great distances and in numerous directions [21]. The colony recruits a higher number of bees to visit floral regions with more flower pollen and nectar. This then brought to the hive with less discomfort than areas having lesser nectar and pollen, to make the best use of its members. The bees in the scout are dispatched from the colony to seek and then assess possible flowers around them which is the first step in their food search process. Among the colony, the fraction of the bees are kept as scout bees over a picking period, and these bees incessantly and haphazardly travel to the floral sections surrounding the hive. The "waggle dance" is a unique dance that these scout bees conduct to let the rest of the bees in the group know about their discovery if they found a flower region that has more food than a set amount. This dance is conducted to relay three crucial pieces of information about the flower locations around the hive and this is the fundamental form of communication amongst the bee colony members. The three factors are the flower patch's orientation, its proximity to the hive, and the caliber of the food available there.

The remaining chapter is organized as follows: the next section deals with the design of a LAA and its mathematical relations, the third talks about the optimization algorithm that is sooty tern optimization (STO) and its mathematical model, fourth dealt in the synthesis of an array, fifth highlights all the obtained results and discussion and the last is the concluding section.

7.2 DESIGN EQUATION OF LAA

An LAA consists of a $2N$ number of isotropic elements. These antennas are laid down in a symmetrical fashion having symmetry with the x-axis as shown in Figure 7.1. In the azimuthal plane, the antenna array factor is decided by the following equation:

$$AF(\theta) = 2\sum_{n=1}^{N} I_n \cos\left(kx_n \cos(\theta) + \varnothing_n\right) \tag{7.1}$$

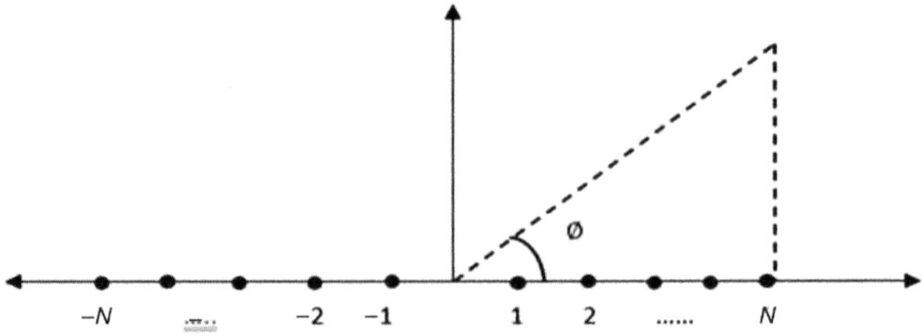

Figure 7.1 Geometry of linear antenna array with 2*N* number of elements placed along the *x*-axis.

where

I_n = Excitation current amplitude
$\varnothing_n$ = Phase of nth element
x_n = Position of nth element in the array
$k = \frac{2\pi}{\lambda}$ is the wave number
θ = Azimuthal angle

For the excitation current amplitude optimization, a certain consideration is that the distance between the elements is taken to be half the wavelength that is $\frac{\lambda}{2}$ and constant phase excitation is applied which is $\varnothing_n = 0$. Keeping these values, the array factor is modified as in the following equation:

$$AF(\theta) = 2\sum_{n=1}^{N} I_n \cos\left(kx_n \cos(\theta)\right)$$

$$AF(\theta) = 2\sum_{n=1}^{N} I_n \cos\left(\frac{2\pi}{\lambda} * \frac{\lambda}{2} \cos(\theta)\right) \tag{7.2}$$

$$AF(\theta) = 2\sum_{n=1}^{N} I_n \cos\left(\pi^* \cos(\theta)\right)$$

7.3 STO ALGORITHM

This bio-inspired metaheuristic algorithm mimics the behavior of a sea bird commonly called sooty tern [22]. Their specific migration and attacking behavior are utilized to fulfill the exploitation and exploration mechanism for a given search space. Large varieties of these species are found in nature having variations in shapes and sizes. They belong to an omnivorous class and feed insects, earthworms, fish, etc. A rain-like effect is created by them using their feet which then attracts the earthworms hidden under the ground. These kinds of birds have their own set of intelligence for finding and catching prey. On a seasonal basis, they migrate from one location to another for the sake of finding an adequate amount of food items that satisfy their needs. This is defined as:

- In their journey they travel in a group and for collision avoidance they set their initial position which will vary for every bird.
- The fittest survival bird must be followed by other members of the group.
- Other members will upgrade their location depending on the fittest member of the family.

To achieve the attack on the target, they follow the spiral-shaped pattern on their journey in space. These characteristics of the bird are framed in such a way that the pattern can be used to optimize the cost function. The process flow chart of STO algorithm is shown in Figure 7.2.

Below is a discussion of the mathematical model used for this method which is primarily focused on migrating and attacking behavior of the bird. The pseudo-code of STO algorithm is outlined in Table 7.1.

7.3.1 Migration behavior

Three conditions are satisfied by the sooty tern during their migration period which are discussed here:

- Collision avoidance

To avoid the collision between its nearby search agents (i.e., sooty terns), S_A is employed to compute the position of the new search agent.

$$\vec{C}_{st} = S_A \times \vec{P}_{st}(z) \tag{7.3}$$

where

$\vec{C}_{st}$ = Position of the agents which have not collided with other agents

$\vec{P}_{st}$ = Search agent present position

z = Iteration used

S_A = In a given search arena, the movement of the agent

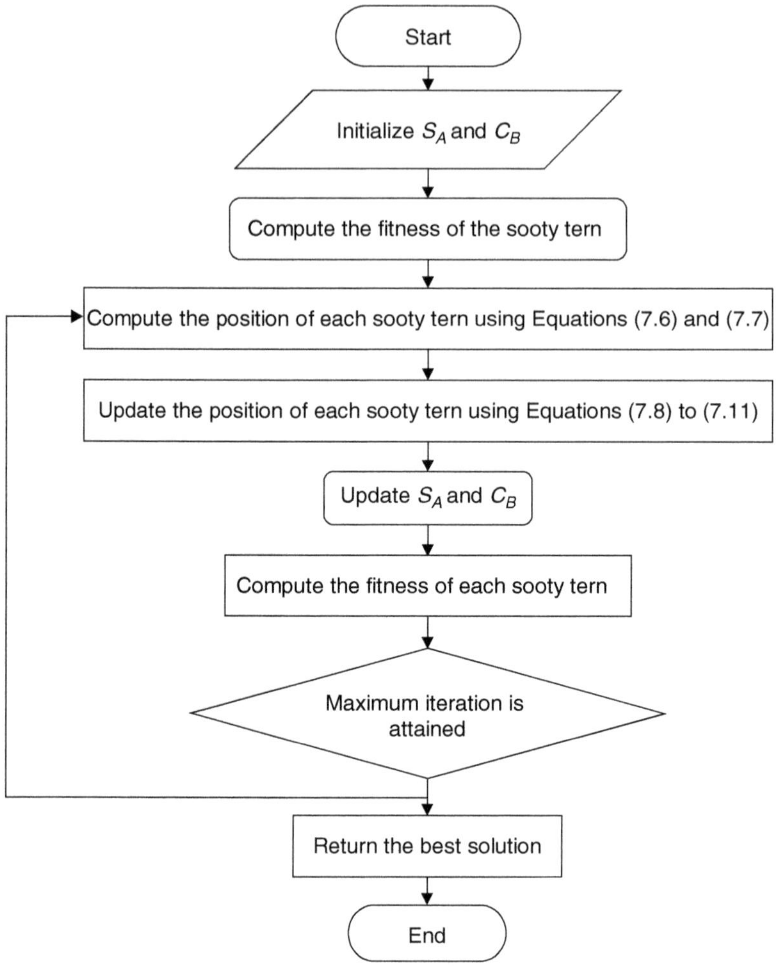

Figure 7.2 Flowchart of the sooty tern algorithm.

$$S_A = C_f - \left(z \times \left(\frac{C_f}{Max_{iterations}} \right) \right) \tag{7.4}$$

where

$z = 0, 1, 2, \ldots, Max_{iterations}$

C_f = Controlling parameter for adjusting S_A which decreases linearly from C_f to 0.

Table 7.1 STO algorithm

Input: $\vec{P}_{st}$ Search agent present position
Output: $\vec{P}_{bst}$ Most qualified search agent
Procedure STOA
S_A and C_B are initialized
Fitness of each search agent is calculated
$\vec{P}_{bst} \longleftarrow$ Most qualified search agent
  while (z < *Maximum Iteration*) **do**
    for every search agent **do**
      Positions are updated using Equation (7.12)
    end for
    S_A and C_B are updated
    Cost of each search agent is computed
    If an appropriate solution exists than the prior optimal, update $\vec{P}_{bst}$.
    z $\longleftarrow$ z + 1
    end while
Return the value of $\vec{P}_{bst}$
Procedure ended

- Converging towards the best neighbor direction

The agents, after avoiding collisions move in the direction of the best neighbor.

$$\vec{M}_{st} = C_B \times \left(\vec{P}_{bst}(z) - \vec{P}_{st}(z)\right) \tag{7.5}$$

where
$\vec{M}_{st}$ indicates the different locations of $\vec{P}_{st}$ for the most qualified search agent $\vec{P}_{bst}$.
C_B = improved exploration is made possible by a random variable.

$$C_B = 0.5 \times R_{and} \tag{7.6}$$

where,
R_{and} = a random number of range between [0, 1]

- Best search agent is updated

The agent or sooty tern can finally update its ranking in terms of the best search agent.

$$\vec{D}_{st} = \vec{C}_{st} + \vec{M}_{st} \tag{7.7}$$

$\vec{D}_{st}$ = gap between best fittest agent and search agent

7.3.2 Attacking behavior

Sooty terns can alter the speed as well as the attack angle during their migration period. Sooty uplifts by using their wings. They produce the spiral kind of behavior in their fly while attacking the prey, as presented:

$$x' = R_{adius} \times \sin(i) \tag{7.8}$$

$$y' = R_{adius} \times \cos(i) \tag{7.9}$$

$$z' = R_{adius} \times i \tag{7.10}$$

$$r = u \times e^{kv} \tag{7.11}$$

where R_{adius} = spiral's each turn radius
i = the variable falls within the range $0 \leq k \leq 2\pi$
u and v = constants define the spiral shape
e = the natural logarithm's base

The values of the constants u and v are 1. Consequently, the new position of the agent is determined using Equation 7.8–7.11:

$$\vec{P}_{st}(z) = \left(\vec{D}_{st} \times (x' + y' + z')\right) \times \vec{P}_{bst}(z) \tag{7.12}$$

where $\vec{P}_{st}(z)$ = update the search agent's position and the best ideal solution is saved.

7.4 SYNTHESIS OF OPTIMIZED ARRAY USING STO ALGORITHM

For the sake of obtaining a better radiation pattern, the amplitude of the excitation current is optimized. Four objective functions are taken into consideration for which the cost functions are formulated, which are stated here:

7.4.1 Peak SLL minimization

Side lobes are the remaining lobes in an antenna's radiation pattern besides the major lobe. These lobes may offer interference and hence are treated as unwanted radiation. Generally, an antenna focuses its energy on a defined direction and to lower the energy wastage minimization of the peak SLL is required for which the fitness function is as follows:

$$fitness = min\left[max\left(\frac{20\log|AF(\theta)|}{max|AF(\theta)|}\right)\right] \tag{7.13}$$

Here, θ is the side lobe region. The target is to minimize the maximum SLL in a defined direction.

7.4.2 Minimization of the close-in SLL

In certain instances, it is desired to minimize the first SLL closer to the main lobe; this is also referred to as minimization of the close-in SLL. For this, the cost function is:

$$fitness = \left(\alpha_1 max\left\{20\log\left|AF\left(\theta_{AS}\right)\right|\right\} + \alpha_2 max\left\{20\log\left|AF\left(\theta_{NS}\right)\right|\right\}\right) \tag{7.14}$$

Here, θ_{AS} = side lobe region, θ_{NS} = close-in SLL. The weights α_1 and α_2 decides the amount of reduction in SLL.

7.4.3 Side lobe average power minimization

The energy of the transmitting antenna is wasted in the form of side lobes, so it is highly desirable to reduce the average power of the side lobes. To achieve this, the objective function is formulated as mentioned below:

$$fitness = \sum_i \frac{1}{\Delta\theta_i}\int_{\theta_{li}}^{\theta_{ui}} |AF(\theta)|^2\, d\theta \tag{7.15}$$

where θ_{li} and θ_{ui} are the lower and upper bounds of the spatial region, respectively, in which the SLL is suppressed, $\Delta\theta_i = \left[\theta_{ui} - \theta_{li}\right]$ is the difference of upper and lower bound values.

7.4.4 SLL minimization along with null placement

By orienting the nulls toward the interfering signal while maintaining the main beam's direction toward the desired signal, the signal reception of the LAA is highly enhanced. To achieve the multi-objective function of side lobe reduction and null placement the following function is formulated:

$$fitness = \sum_i \frac{1}{\Delta\theta_i}\int_{\theta_{li}}^{\theta_{ui}} |AF(\theta)|^2\, d\theta + \sum_k \left|AF\left(\theta_k\right)\right|^2 \tag{7.16}$$

7.5 RESULTS AND DISCUSSION

In this study, the amplitude of the 10-element and 14-element LAA is optimized by using various metaheuristic algorithms. Based on the statistical analysis, it is found that the STO algorithm offers appropriate results as compared to the other applied algorithms for this problem.

The STO is implemented on MATLAB and the execution was repeated 10 times. In all cases, the size of the population is set equal to 50 and the maximum value of iterations for each run is taken to be 1000. The optimization of the current amplitude is performed here for the sake of obtaining the desired pattern which offers peak SLL minimization, close-in SLL reduction, side lobe average power minimization, and SLL minimization along with null placement. To achieve these goals for improving the radiation characteristics of a LAA, four design examples A, B, C, and D are discussed here while their details are outlined in Table 7.2.

Design example A illustrates the application of STO in optimizing the current amplitudes of the 10-element LAA for achieving the peak SLL minimization. For the analysis of the array factor with respect to the azimuth angle of the LAA from 0° to 180°, Equation 7.2 is used here. The SLL minimization is performed in the regions [0°,76°] and [104°,180°]. The fitness function used by STO for this design example is provided in Equation 7.13.

The current amplitudes are optimized using the ALO, Dragonfly Algorithm (DA), Harris Halk Optimization (HHO), and STO algorithms. The obtained values of the current amplitudes are provided in Table 7.3. This table indicates the current amplitude of the positive half of the LAA as the negative half of the antenna elements has the same set of excitation current amplitude in the case of a uniform LAA.

The antenna array pattern of this design is shown in Figure 7.3 which also indicates the array pattern of uniform array and the pattern of 10-element LAA obtained by using various optimization algorithms. In this, it is seen that the array factor obtained by applying various optimization algorithms finds the best value with STO.

The first SLL of the uniform LAA is –12.97 dB, while the first SLL obtained by applying the various algorithms is much less the lowest one

Table 7.2 Details of the design example

Design example	*Objective*	*Number of antenna elements in a LAA*
A	Peak side lobe level minimization	10
B	Minimization of the close-in SLL	10
C	Side lobe average power minimization	14
D	SLL minimization along with null placement	14

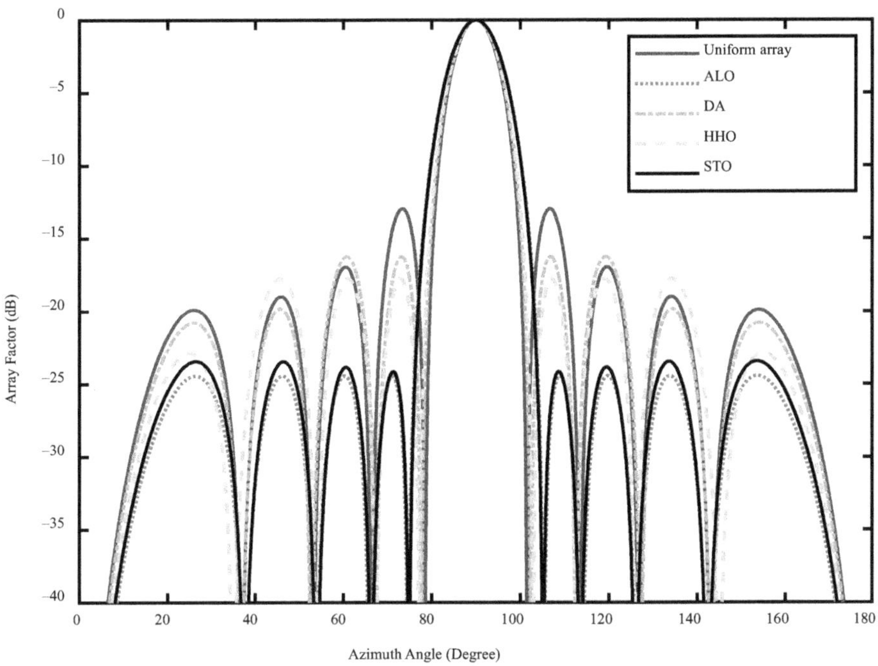

Figure 7.3 Antenna array pattern for design example A.

Table 7.3 Obtained amplitudes of the positive part of the 10-element LAA using different optimization methods for design example A

Method			*Optimized current amplitudes*		
ALO	1.0000	0.9018	0.7276	0.5150	0.4159]
DA	1.0000	0.8715	0.7740	0.7740	0.7740]
HHO	1.0000	0.9899	0.7385	0.7385	0.7385]
STO	1.0000	0.9058	0.7320	0.5171	0.4452]

obtained by STO of about –24.18 dB, which is 11.21 dB down with respect to the uniform array.

The ALO shows the consistent value of all the SLLs, but the best result is obtained by applying the STO algorithm which offers the best-optimized values of excitation current amplitude that is most effective in the estimation of the array factor of the antenna array. The convergence curve of all the algorithms to achieve the peak SLL minimization is shown in Figure 7.4.

Using these algorithms, the SLLs are obtained in decibels which is mentioned in Table 7.4. The standard deviation in DA is very high.

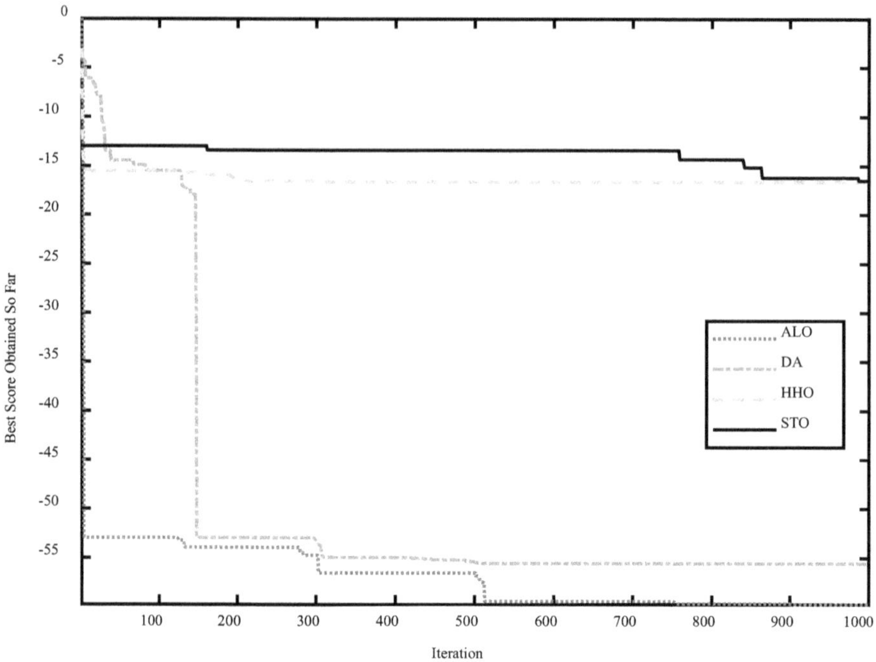

Figure 7.4 Convergence curve for design example A.

Table 7.4 Statistical values of SLL and processing time of ALO, DA, HHO, and STO for design example A

	Uniform array	*ALO*	*DA*	*HHO*	*STO*
SLLs	–12.97	–24.43	–20.79	–22.88	–23.44
	–16.95	–24.46	–19.86	–17.76	–23.47
	–18.99	–24.47	–16.26	–17.76	–23.86
	–19.90	–24.15	–16.29	–17.75	–24.21
Best SLL (dB)	–19.90	–24.47	–20.79	–22.88	–24.21
Mean SLL (dB)	–17.20	–24.37	–18.30	–19.03	–23.74
Worst SLL (dB)	–12.97	–24.15	–16.26	–17.75	–23.44
Std. Dev. SLL (dB)	2.6669	0.1321	2.0515	2.2184	0.3154
Best CPU Time (s)	—	62	108	34	18

Comparing all these values of the SLL, the SLL obtained by applying STO has the lowest value of –24.21 for best SLL and its mean SLL is also better as compared with other state-of-the-art optimization algorithms which is applied here in this problem.

The computational time to find out the optimized weights by STO also has the lowest value of 18 seconds and the highest time for finding the optimal

Table 7.5 Amplitudes of the positive portion of the 10-element LAA using different methods for design example B

Method	*Optimized current amplitudes*				
ALO	1.0000	0.6877	0.5111	0.5111	0.5147]
DA	1.0000	0.8754	0.6290	0.4298	0.6135]
HHO	1.0000	0.5075	0.5057	0.5057	0.5057]
STO	1.0000	0.7013	0.5147	0.5122	0.5163]

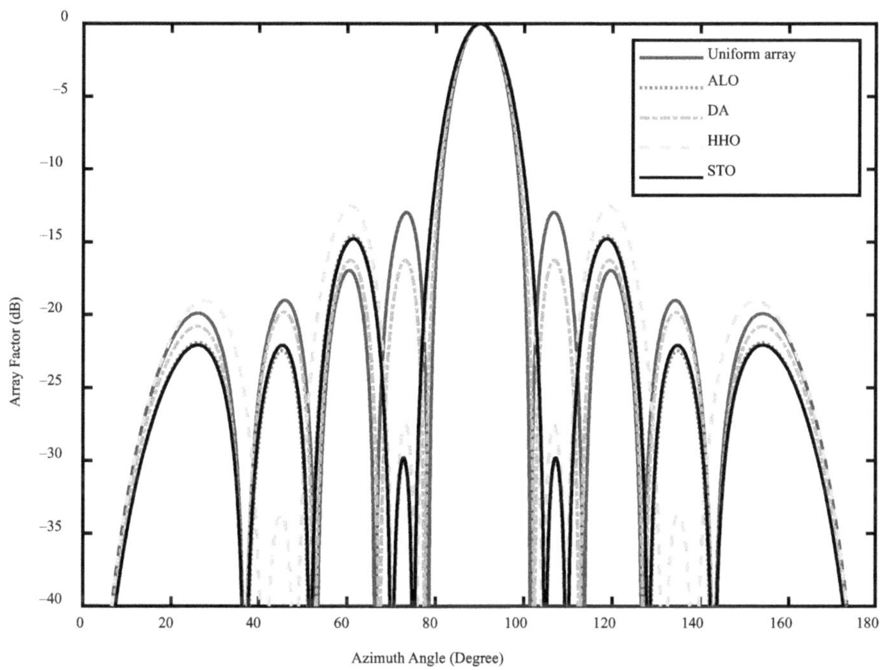

Figure 7.5 Antenna array pattern for design example B.

weights of the antenna elements is taken by the DA of about 108 seconds. So, for this design example, an STO algorithm produces the optimal values of the excitation current amplitude which offers the minimization of peak values of the SLLs.

The design example B is a 10-element LAA which is considered for close-in SLL reduction. The side lobes lie in the region [0°,76°] and [104°,180°], out of which the close-in or first side lobe region extends from [69°,76°] and [104°,111°], which is now considered for minimization. For this purpose, the cost function formulated is mentioned in Equation 7.14. It is then optimized by using the said algorithms and the optimized current amplitudes are provided in Table 7.5.

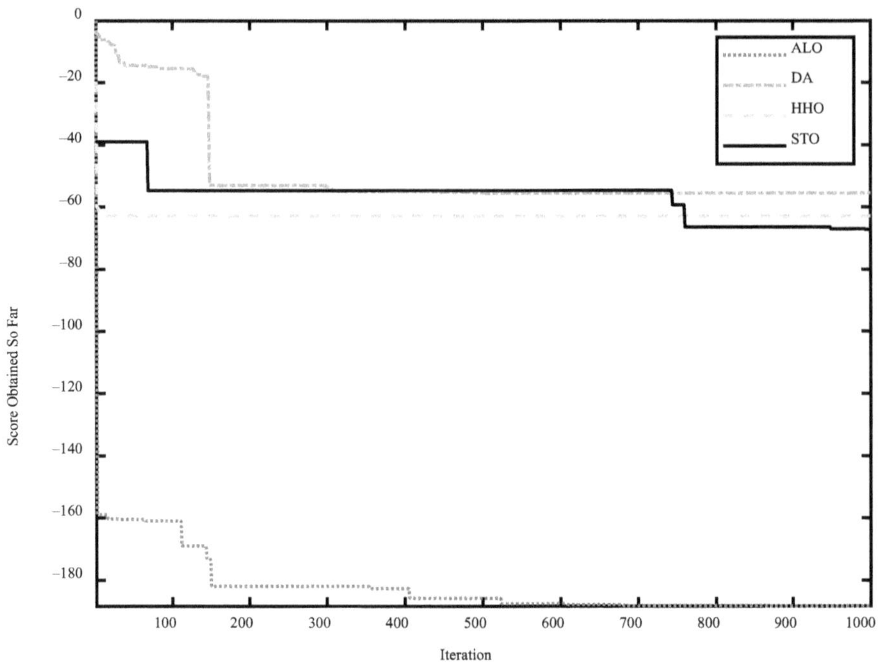

Figure 7.6 Convergence curve for design example B.

With this set of weights for current amplitude, the radiation pattern for close-in SLL reduction is shown in Figure 7.5 which clearly indicates that the first SLL is reduced up to –29.84 dB by using the STO algorithm which is 16.87 dB less in comparison with the first side lobe of the uniform LAA.

STO achieves the lowest value when compared to other optimization algorithms that have been used here in this design example. Different algorithms take different time for convergence, here STO takes 76 seconds to reach the optimum solution. HHO takes the largest time for its convergence. The convergence curve for close-in SLL reduction is shown in Figure 7.6; in this, the STO achieves the optimal weights too quickly in comparison to the other applied optimization algorithm.

The design example C is a 14-element LAA which is designed to meet the side lobe average power minimization. The objective function formulated to achieve the side lobe average power minimization is mentioned in Equation 7.15. The array factor is computed by using Equation 7.2 and is shown in Figure 7.7.

The four algorithms are applied for this design to optimize the current amplitude and the STO outperforms the rest of the applied algorithms. The side lobe average power is reduced, the comparison of the SLLs is mentioned in Table 7.6.

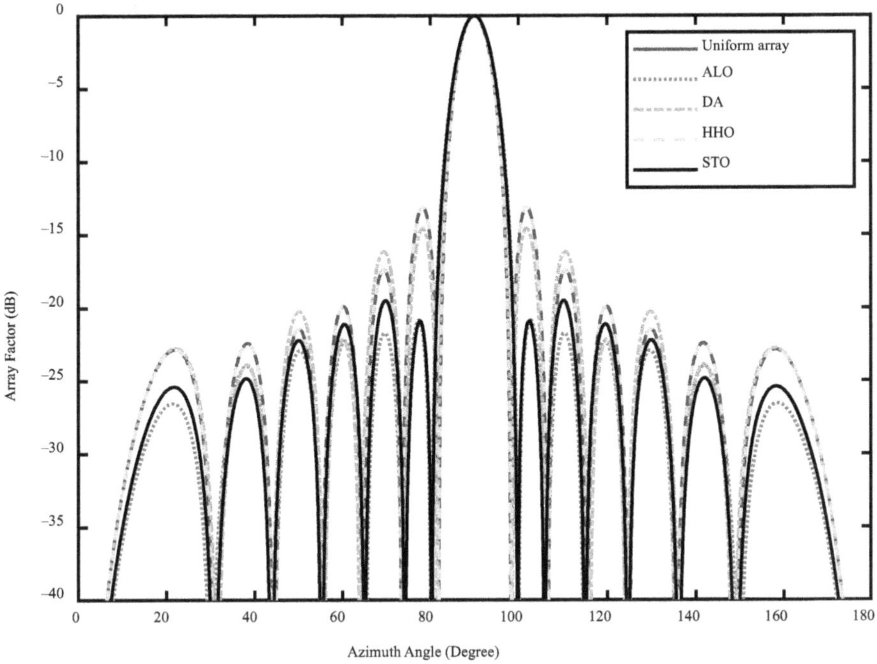

Figure 7.7 Antenna array pattern for design example C.

Table 7.6 SLLs of design example C

Method	*SLLs (dB)*					
	I	*II*	*III*	*IV*	*V*	*VI*
Uniform Array	−13.12	−17.44	−19.89	−21.49	−22.42	−22.87
ALO	−20.84	−21.79	−22.22	−22.76	−23.88	−26.55
DA	−14.57	−16.16	−22.19	−20.24	−23.89	−22.87
HHO	−13.11	−17.41	−19.89	−21.47	−22.44	−22.87
STO	−20.92	−19.54	−21.18	−22.27	−24.83	−25.42

Table 7.7 Amplitude of the positive region of the 10-element LAA using different optimization techniques for design example C

Method	*Optimized current amplitudes*						
ALO	1.0000	0.9645	0.8739	0.7872	0.6179	0.5708	0.5708]
DA	1.0000	0.8286	0.8274	0.8274	0.8274	0.8285	0.8274]
HHO	1.0000	0.7541	0.7312	0.6125	0.6024	0.5821	0.5612]
STO	1.0000	0.9278	0.8493	0.7253	0.6031	0.6017	0.6017]

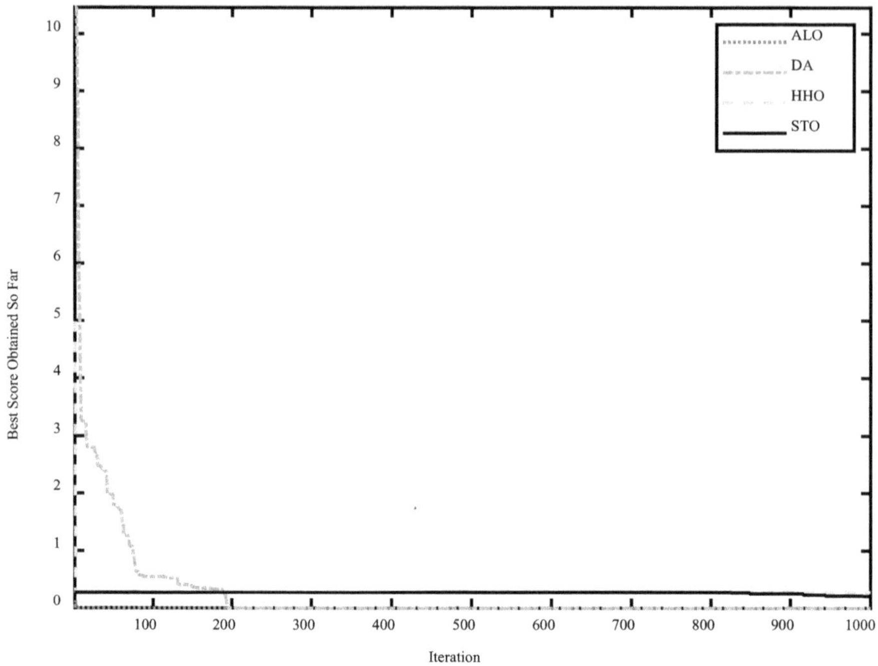

Figure 7.8 Convergence profile for design example C.

The optimized values of the excitation current amplitudes for design example C are shown in Table 7.7. Using the values obtained for current amplitude the array factor is computed. The convergence curve for the different algorithms is shown in Figure 7.8.

The design example D is presented, which is a 14-element LAA; in this, the multiobjective function is implemented for SLL reduction along with the placement of nulls in the direction of the interfering signal. By directing the nulls in the direction of the interfering signal while maintaining the main beam toward the direction of the desired signal, the linear array's ability to receive the desired signal is improved to a greater extent. The investigation of array pattern nulling approaches has been spurred by the rising electromagnetic pollution. The undesired direction or direction of the interfering signal is decided as $\theta = 64°, 76°, 104°$, and $116°$. The cost function defined for this design example is mentioned in Equation 7.16. The obtained variation of the array factor is plotted in Figure 7.9 and the convergence curve is shown in Figure 7.10. The obtained current amplitude is provided in Table 7.8.

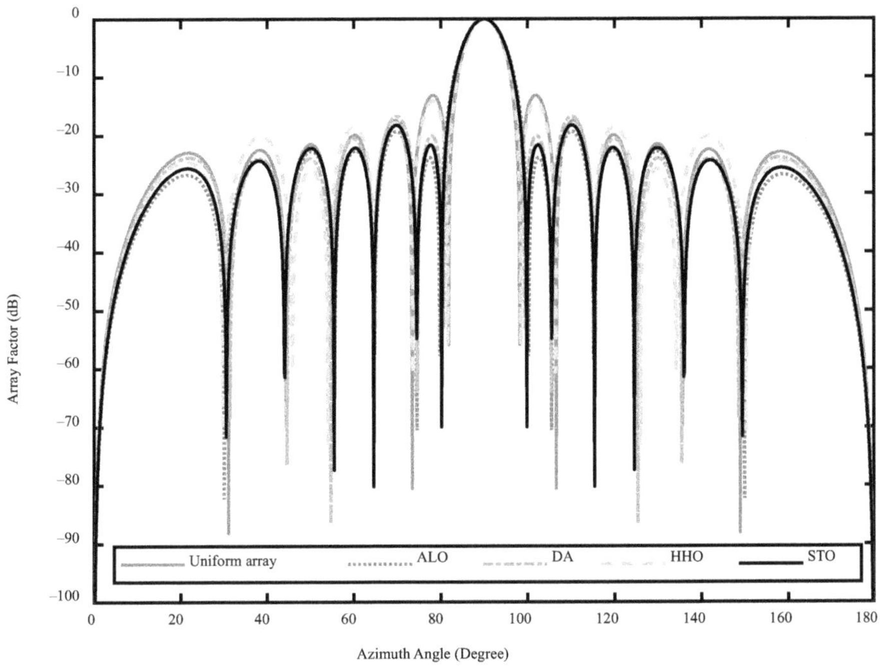

Figure 7.9 The array pattern for design example D.

Table 7.8 Amplitudes of the14-element LAA for design example D

Method	*Optimized current amplitudes*						
ALO	1.0000	0.9100	0.7760	0.6875	0.5493	0.5361	0.5361]
DA	1.0000	0.9261	0.7798	0.6517	0.6517	0.6517	0.6517]
HHO	1.0000	0.8221	0.8221	0.8221	0.8221	0.8221	0.8221]
STO	1.0000	0.8904	0.7781	0.6881	0.5845	0.5845	0.5845]

7.6 CONCLUSION

This chapter has introduced the STO algorithm for the antenna domain. Here, STO is used for the pattern synthesis of a LAA. This is applied to obtain the excitation current amplitude of the elements of the LAA to achieve the minimization in SLLs, reduction of first SLL, and positioning of the nulls in the desired direction. The results so obtained are compared with a uniform LAA which is unoptimized and with the antenna arrays that are optimized by using metaheuristic algorithms like ALO, DA, HHO, and STO. The obtained optimized solution indicates that STO offers improved

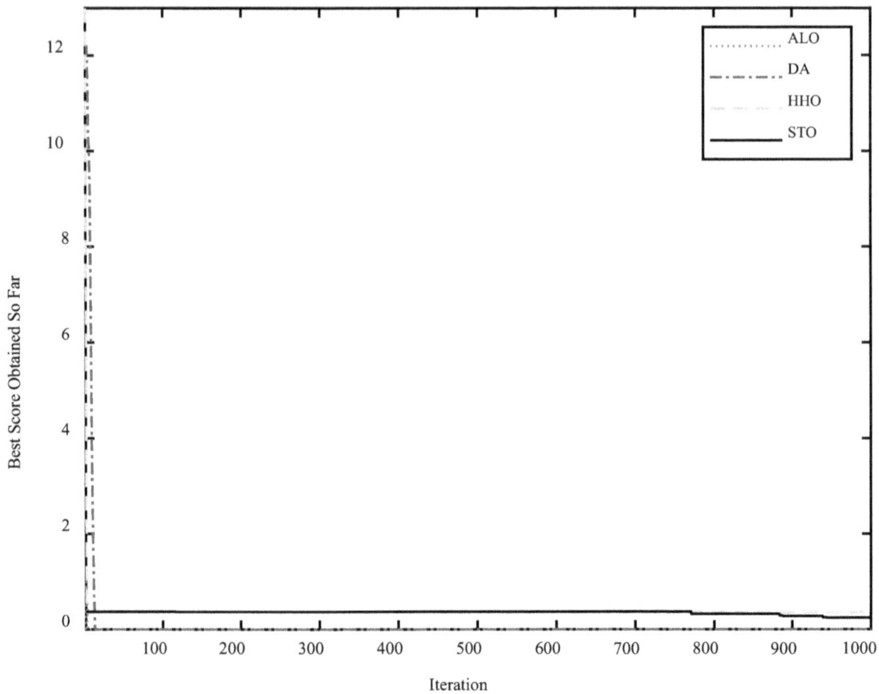

Figure 7.10 Convergence curve for design example D.

performance as compared to the uniform array and with another optimization algorithm.

This suggests that STO can effectively be used for the antenna array optimization for obtaining the optimal solution and hence is well suited for designing an antenna array.

REFERENCES

[1] A. Durmus, R. Kurban, and E. Karakose, "A comparison of swarm-based optimization algorithms in linear antenna array synthesis," *Journal of Computational Electronics*, vol. 20, no. 4, pp. 1520–1531, 2021.

[2] B. Goswami and D. Mandal, "A genetic algorithm for the level control of nulls and side lobes in linear antenna arrays," *Journal of King Saud University—Computer and Information Sciences*, vol. 25, no. 2, pp. 117–126, 2013.

[3] K. Guney and M. Onay, "Amplitude-only pattern nulling of linear antenna arrays with the use of bees algorithm," *Progress in Electromagnetics Research—PIER*, vol. 70, pp. 21–36, 2007.

[4] C. Zhang, X. Fu, L. P. Ligthart, S. Peng, and M. Xie, "Synthesis of broadside linear aperiodic arrays with sidelobe suppression and null steering using whale optimization algorithm," *IEEE Antennas and Wireless Propagation Letters,* vol. 17, no. 2, pp. 347–350, 2018.

[5] A. Sharma, "Antenna array pattern synthesis using metaheuristic algorithms: A review," *IETE Technical Review*, vol. 40, no. 1, pp. 90–115, 2023.

[6] S. K. Goudos, V. Moysiadou, T. Samaras, K. Siakavara, and J. N. Sahalos, "Application of a comprehensive learning particle swarm optimizer to unequally spaced linear array synthesis with sidelobe level suppression and null control," *IEEE Antennas and Wireless Propagation Letters*, vol. 9, pp. 125–129, 2010.

[7] T. Pavani, K. Padmavathi, Ch. U. Kumari, and A. Ushasree, "Design of array antennas via atom search optimization," *Materials Today: Proceedings*, vol. 80, pp. 2051–2054, 2023.

[8] U. Singh and M. Rattan, "Design of linear and circular antenna arrays using cuckoo optimization algorithm," *Progress in Electromagnetics research C*, vol. 46, pp. 1–11, 2014.

[9] L. Pappula and D. Ghosh, "Linear antenna array synthesis using cat swarm optimization," *AEU—International Journal of Electronics and Communications*, vol. 68, no. 6, pp. 540–549, 2014.

[10] W. C. Weng, F. Yang, and A. Z. Elsherbeni, "Linear antenna array synthesis using Taguchi's method: A novel optimization technique in electromagnetics," *IEEE Trans Antennas Propagation*, vol. 55, no. 3, pp. 723–730, 2007.

[11] A. Das, D. Mandal, S. P. Ghoshal, and R. Kar, "Moth flame optimization-based design of linear and circular antenna array for side lobe reduction," *International Journal of Numerical Modelling: Electronic Networks, Devices and Fields*, vol. 32, no. 1, pp. 1–15, 2019.

[12] L. H. Abderrahmane and B. Boussouar, "New optimisation algorithm for planar antenna array synthesis," *AEU—International Journal of Electronics and Communications*, vol. 66, no. 9, pp. 752–757, 2012.

[13] X. Li and M. Yin, "Optimal synthesis of linear antenna array with composite differential evolution algorithm," *Scientia Iranica*, vol. 19, no. 6, pp. 1780–1787, 2012.

[14] S. I. Farghaly, H. E. Seleem, M. M. Abd-Elnaby, and A. H. Hussein, "Pencil and shaped beam patterns synthesis using a hybrid GA/l1 optimization and its application to improve spectral efficiency of massive MIMO systems," *IEEE Access*, vol. 9, pp. 38202–38220, 2021.

[15] P. Saxena and A. Kothari, "Ant Lion Optimization algorithm to control side lobe level and null depths in linear antenna arrays," *AEU—International Journal of Electronics and Communications*, vol. 70, no. 9, pp. 1339–1349, 2016.

[16] P. Saxena and A. Kothari, "Optimal pattern synthesis of linear antenna array using grey wolf optimization algorithm," *International Journal of Antennas and Propagation*, vol. 2016, no. 1, pp. 1–11, 2016.

[17] K. R. Subhashini, "Runner-root algorithm to control sidelobe level and null depths in linear antenna arrays," *Arabian Journal for Science and Engineering*, vol. 45, no. 3, pp. 1513–1529, 2020.
[18] F. Güneş and F. Tokan, "Pattern Search optimization with applications on synthesis of linear antenna arrays," *Expert Systems with Applications*, vol. 37, no. 6, pp. 4698–4705, 2010.
[19] N. Singh Grewal, M. Rattan, and M. Singh Patterh, "A linear antenna array failure correction using improved bat algorithm," *International Journal of RF and Microwave Computer-Aided Engineering*, vol. 27, no. 7, 2017.
[20] U. Singh, H. Kumar, and T. S. Kamal, "Linear array synthesis using biogeography-based optimization," *Progress in Electromagnetics Research M*, vol. 11, pp. 25–36, 2010.
[21] K. Guney and M. Onay, "Bees algorithm for interference suppression of linear antenna arrays by controlling the phase-only and both the amplitude and phase," *Expert Systems with Applications*, vol. 37, no. 4, pp. 3129–3135, 2010.
[22] G. Dhiman and A. Kaur, "STOA: A bio-inspired based optimization algorithm for industrial engineering problems," *Engineering Applications of Artificial Intelligence,* vol. 82, pp. 148–174, 2019.

Chapter 8

1-D and 2-D beam-switching transmitarray antennas for millimeter wave base stations and vehicular networks

Javid Ahmad Ganie and Kushmanda Saurav

8.1 INTRODUCTION

Millimeter-wave frequencies, envisioned for future 5G communication, offer superior channel capacity compared to current wireless networks [1]. However, electromagnetic waves encounter increased propagation loss at these frequencies. This obstacle is addressed by utilizing high-gain antennas, which offset the loss and improve the signal to noise ratio (SNR) [2]. Nonetheless, the narrow beam width of high-gain antennas makes beam adjustment challenging. Due to the smaller wavelength characteristic of mm-wave frequencies, antenna size decreases, enabling the clustering of multiple antennas to establish a multiple input multiple output (MIMO) system, thereby assisting in beam switching [3].

Array antennas offer a method for crafting antennas having high gain [4], yet directing the main radiation beam toward a specific user arbitrarily requires phased array antennas, that rely on active circuit components for phase control. This dependence may incur losses and diminish the antenna radiation efficiency. Alternative methods of designing antennas with high gain include transmit arrays, lens antennas, and reflect arrays [5, 6]. While lens antennas pose challenges due to their curved geometry, reflect arrays suffer from feed blockage, making transmitarrays the suitable choice for designing high-gain antennas. Researchers are drawn to designing transmitarrays due to their appealing features such as high gain, straightforward feeding technique, affordability, and beam switching capabilities. Broadly, transmitarray elements can be designed using three methods.

The initial approach entails utilizing various unit cells (UCs) with transmission coefficient phases spanning from 0 to 360 degrees [7–12]. The second approach involves linking metallic resonators through a via-based transmission line structured on different dielectric layers. The third approach utilizes UCs with polarization rotating capabilities, featuring various dielectric and metallic layer structures [13–16].

DOI: 10.1201/9781003470854-8

Additionally, the literature covers switchable beam transmitarray antennas that integrate MIMO feeds [17–18]. The transmitarray outlined in reference [17], composed of 2 dielectric layers with an intervening gap, utilizes an aperture-coupled antenna array as its feeding mechanism.

In this chapter, one-dimensional (1-D) and two-dimensional (2-D) switchable beam transmitarrays are presented, showcasing phase approximations of 0 and π. The antennas deliver a maximum gain of 19.3 and 19.7 dBi, respectively. Different beams exhibit 3-dB beam widths ranging from 9.8 to 11.2 degrees. The offset feed elements along both the x- and y-axes produce a beam along ±13 degrees in the case of a 1-D beam-switching antenna, whereas the beam direction for the 2-D antenna is ±15 degrees. These antenna designs are well suited for installation in a variety of vehicles, providing dependable communication with diverse networks.

8.2 1-D BEAM-SWITCHING TRANSMITARRAY DESIGN

8.2.1 Operation of transmitarray and unit cell design

The transmitarray functions in transmission mode, whereby an incident spherical wave from the feed antenna traverses via the panel, resulting in a high-gain radiation beam.

Figures 8.1 and 8.2 display the UCs corresponding to bit 0 and bit 1, respectively, utilized in crafting the transmitarray. These basic UCs are crafted utilizing 2 dielectric layers made of Rogers RT5880, featuring a relative permittivity of 2.2 (1.52 mm thick). The metallic layers, totaling 3, are printed on the 2 dielectric layers, as depicted in Figures 8.1 and 8.2. Figure 8.3 demonstrates that the polarization rotating elements for bit 0 and bit 1 exhibit a band pass transmission response. This band pass response can be represented by transmission lines connecting shunt capacitors and shunt inductors.

8.2.2 Transmitarray configuration

The phase distribution across the transmitarray panel is determined using equations outlined in reference [14] and is implemented through two levels of phases. The transmitarray surface consists of a grid of 29 × 29 phase delay cells. The surface is excited by a patch MIMO antenna, consisting of 3 elements, and works in the frequency range of 29.7–30.7 GHz (Figure 8.4). The MIMO antenna utilizes Rogers RT5880 (0.254 mm) as the dielectric substrate material. The three patch antennas are located at the center, 10 and –10 mm along the x-axis, respectively. Figure 8.5 depicts the general schematic of the transmitarray, whereas Figure 8.6 illustrates the phase

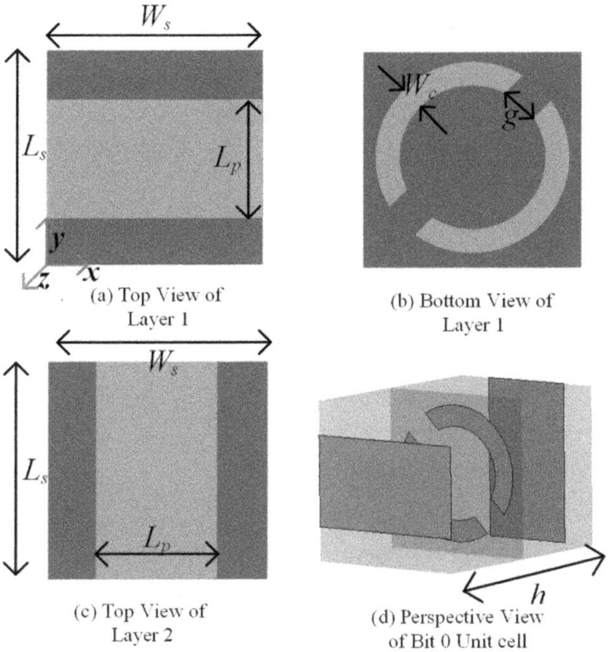

Figure 8.1 Diagram illustrating the design of the bit-0 element for the transmitarray.

distribution at 30.5 GHz for a focal distance, $f = 30$ mm. The effective size of the entire transmitarray, established by the focal distance to aperture diameter (f/d) ratio, is 0.49. Due to the distinct phase distribution seen by the two offset antenna elements on the surface, radiation beams tilted at +13 and −13 degrees are achieved relative to the middle element.

8.2.2.1 Fabrication and measurement

The beam-switching transmitarray presented here is manufactured using a conventional photolithography method. The fabricated prototype is depicted in Figure 8.7. The transmitarray panel consists of two layers, which are firmly joined together using screws located at the four corners of the surface. The assessment of the transmitarray is done inside the measurement facility, using the PNA N5224B.

Figure 8.8 depicts both the measured and simulated *S*-parameters of the MIMO patch antenna. To optimize isolation while minimizing switching loss, the space between elements is kept at λ_0. While increasing the spacing could potentially improve isolation further, it would also result in higher

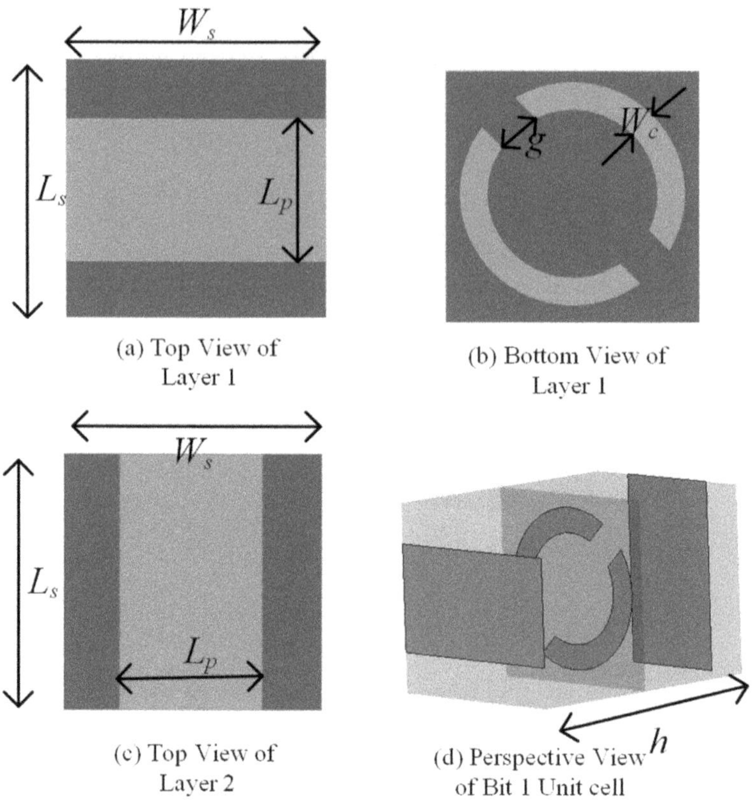

Figure 8.2 Diagram illustrating the design of the bit-0 element for the transmitarray.

switching loss. The observed variance in the return loss in comparison to the simulated one can be associated with fabrication errors. At 30.5 GHz, the transmitarray's radiation patterns are evaluated. Figures 8.9 and 8.10 show both measured and simulated radiation patterns of port 1 in both planes, respectively. Additionally, Figures 8.11 and 8.12 show the patterns of elements 2 and 3 in the *xz* planes. The beam of port 1 exhibits an half power beam width (HPBW) of 10.1 degrees in both the principal planes.

Figure 8.13 presents the measured and simulated gains of ports 1 and 2. A decrease in magnitude is observed in the measured gain compared to the simulated counterpart, attributable to tolerances in the measurement process. The steering loss, defined as the reduction in gain of port 2 with reference to port 1 is approximately 1 dB, falling within an acceptable range.

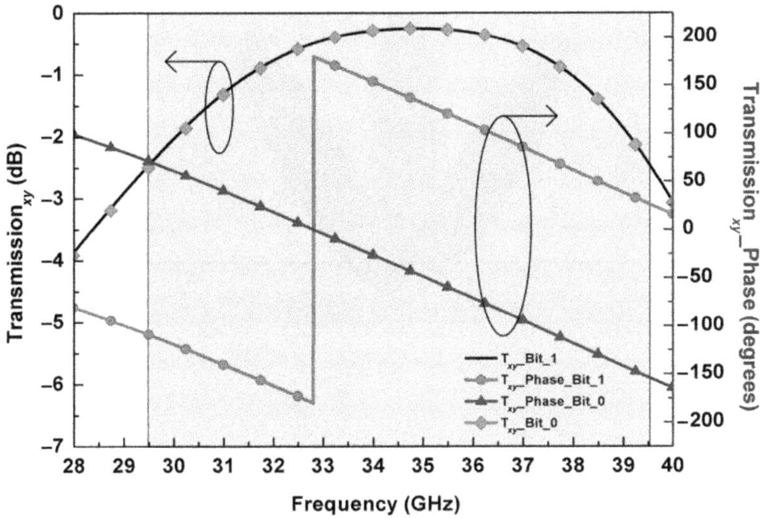

Figure 8.3 Simulated magnitude and phase of the transmission coefficient for the bit-0 and bit-1 of the transmitarray.

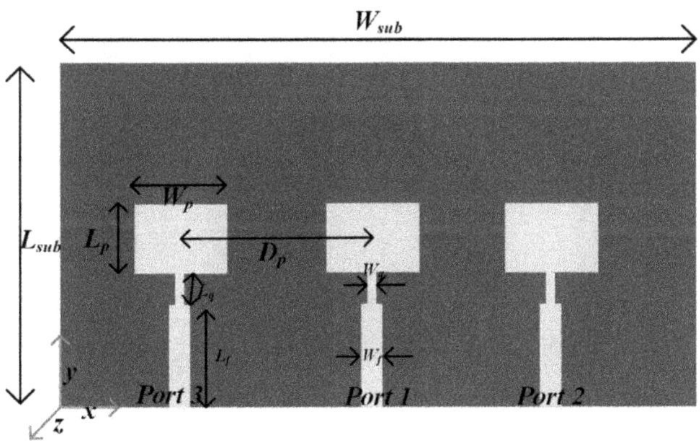

Figure 8.4 Feed MIMO antenna for the 2-D beam switching antenna.

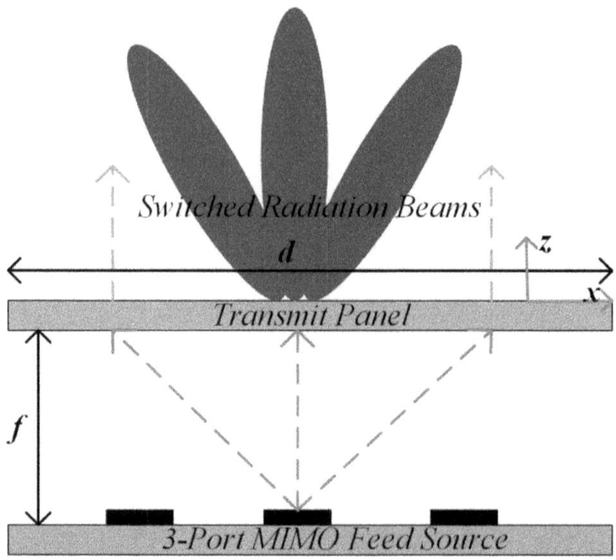

Figure 8.5 Layout of the suggested 2-D beam-switching antenna.

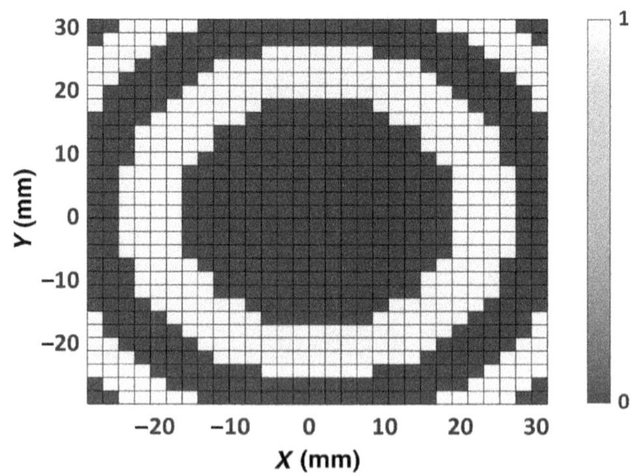

Figure 8.6 Phase distribution of the transmitarray surface at 30.5 GHz with 1-bit resolution.

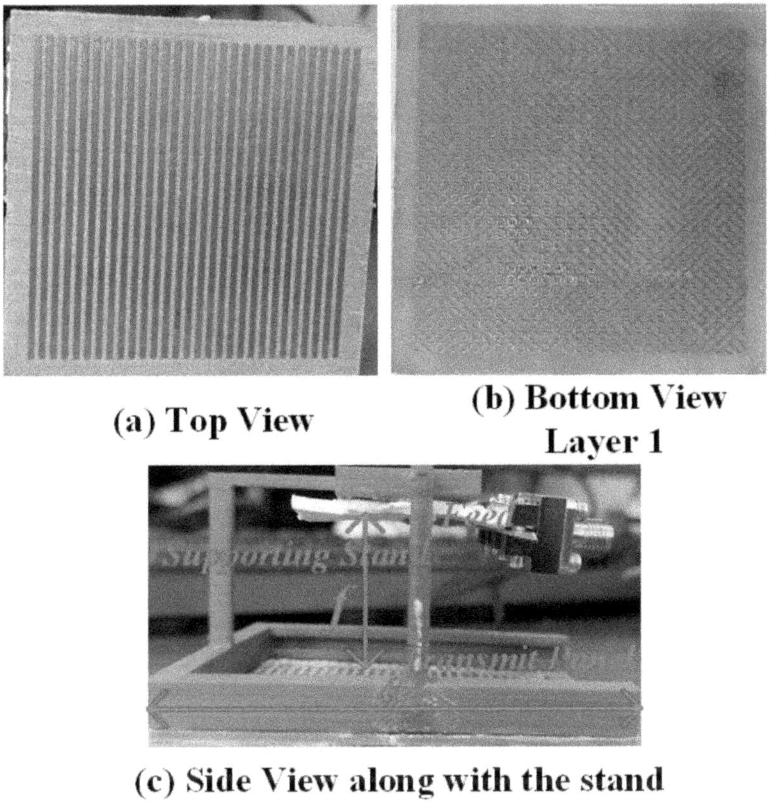

Figure 8.7 Manufactured transmitarray antenna with accompanying support stand.

8.3 2-D BEAM-SWITCHING TRANSMITARRAY DESIGN

8.3.1 Transmitarray element design and simulation

Figure 8.14 shows the phase shift elements for the transmitarray. The element consists of three metallic layers and two dielectric substrate layers (Rogers RT5880).

The upper dielectric layer is clad with metallic layers on both surfaces (top and middle layers), whereas the lower dielectric layer accommodates only a single copper layer (bottom layer). The assembly of these two dielectric substrate layers is achieved with the use of Teflon screws. In terms of the metallic layers, both the upper and lower layers showcase metallic cross structures rotated orthogonally, whereas the middle metallic layer incorporates end-loaded dipole structures rotated at ±45 degrees. Using the frequency domain solver in CST Microwave Studio, the simulation of the

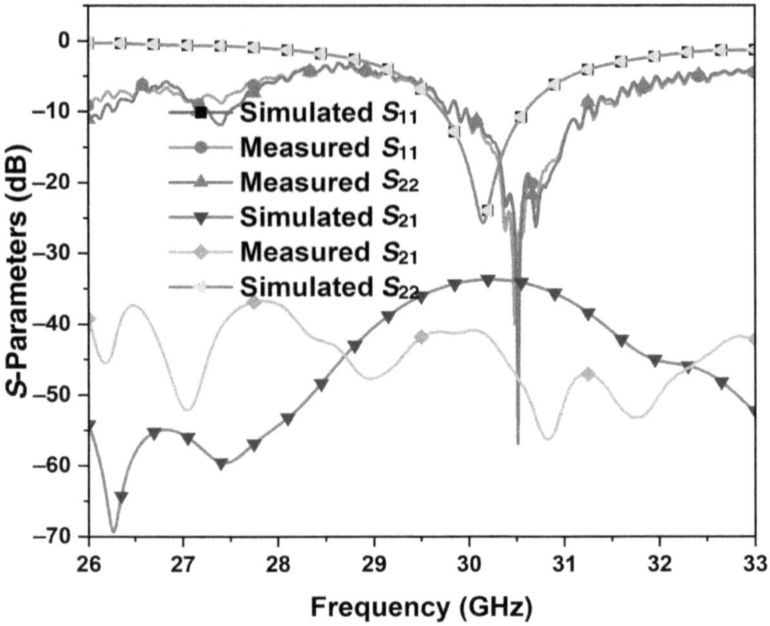

Figure 8.8 Measured and simulated *S*-parameters of the feed elements.

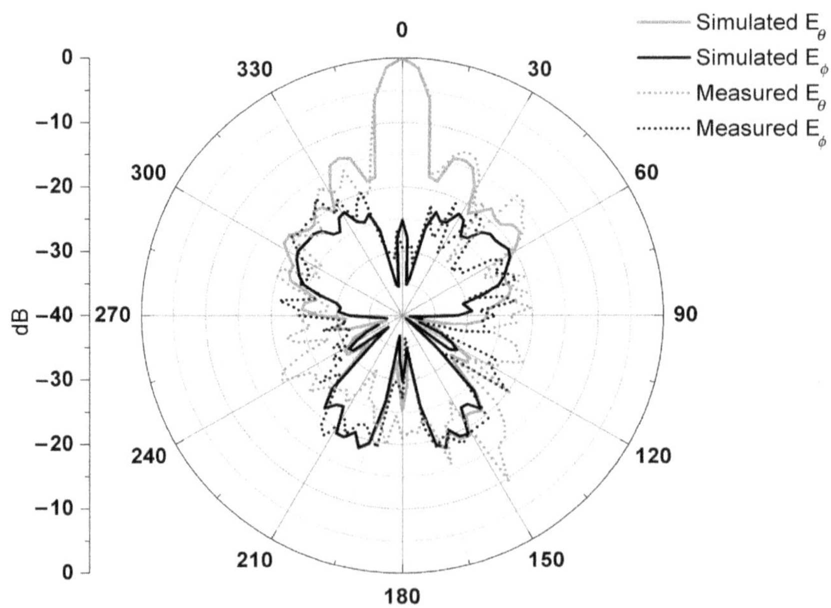

Figure 8.9 Measured and simulated radiation pattern of port 1 at 30.5 GHz in the *xz* plane.

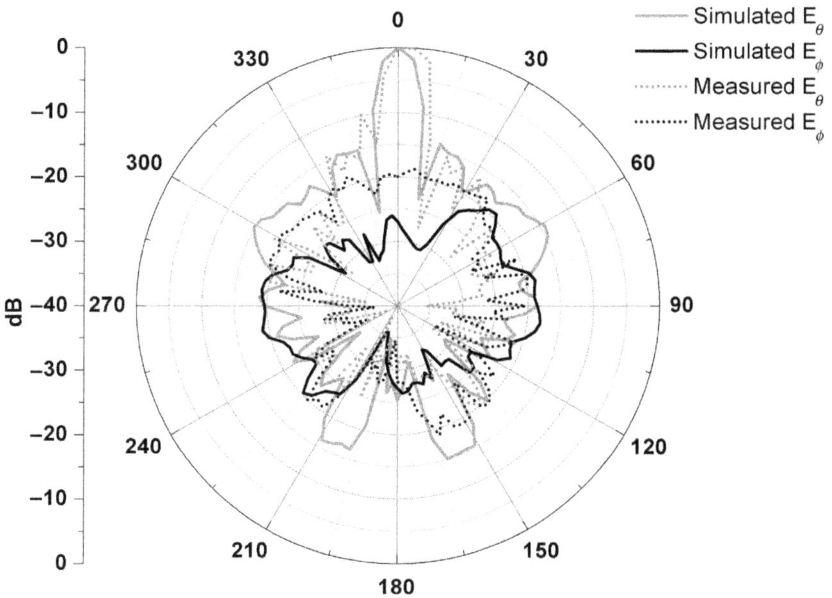

Figure 8.10 Measured and simulated radiation pattern of port 1 at 30.5 GHz in the *yz* plane.

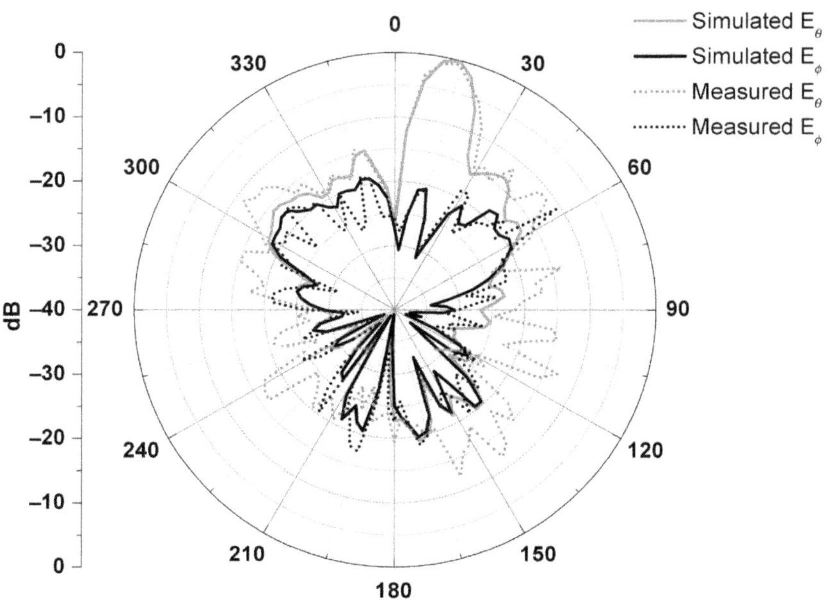

Figure 8.11 Measured and simulated radiation pattern of port 2 at 30.5GHz in *xz* plane.

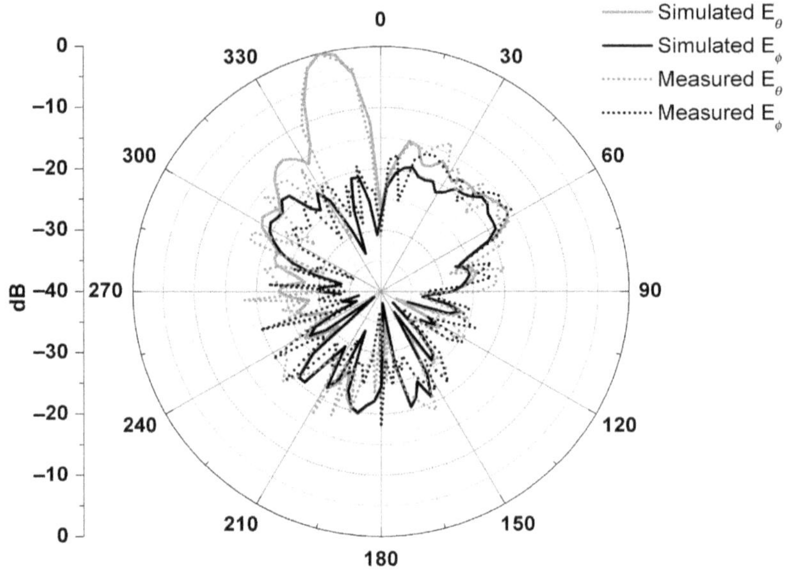

Figure 8.12 Measured and simulated radiation pattern of the port 3 at 30.5GHz in *xz*.

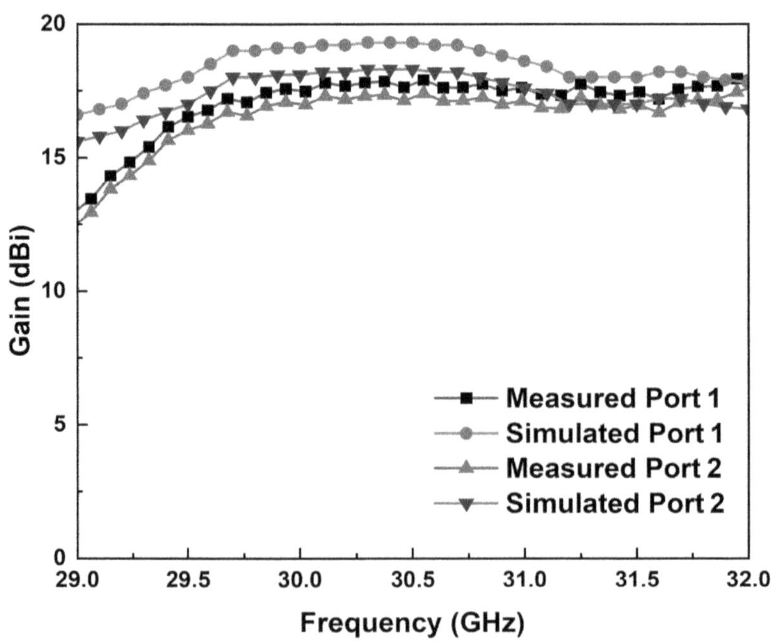

Figure 8.13 Measured and simulated gain of the switchable beam antenna.

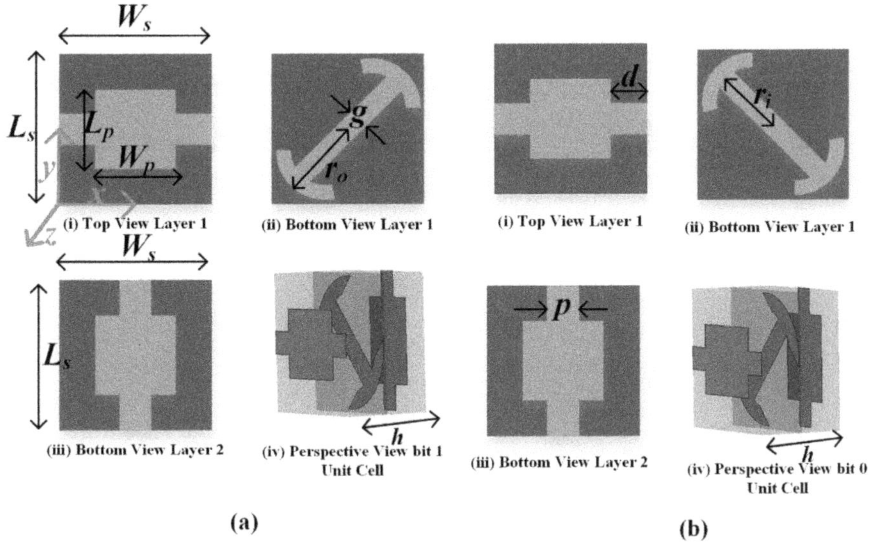

Figure 8.14 Designs of the UCs for the 2-D beam-switching antenna: (a) UC design for bit-0 and (b) UC design for bit-1.

phase shift element involves applying UC boundary conditions. The phase difference of 180 degrees is obtained through the ±45 degrees rotation in the dipole structure. In particular, the wave polarized along the *y*-axis at the input is transformed into a wave polarized along the *x*-axis at the output.

Observations indicate that altering the patch length from 0.6 to 1.2 mm enhances the magnitude of the transmission coefficient within the 24–40 GHz frequency range. Nevertheless, continued increments in patch length lead to a reduction in transmission bandwidth. As a result, a patch length of 1.2 mm is chosen.

Another critical factor affecting the transmission frequency range of the elements is *g*.

As the length (L_p) increases, so does the area of the copper structure. This increment in the 'L_p' of the metal strips results in heightened inductance, consequently leading to a larger '*Z*' of the element as depicted in Figure 8.15. The heightened '*Z*' decreases the magnitude of the reflection, consequently improving the amplitude of the transmission coefficient. The same principle applies to widening the '*g*' of the dipole strip. Expanding the width of the copper strips enhances transmission magnitude up to a specific threshold as shown in Figure 8.16; however, beyond that point, further widening leads to a change in the frequency.

The change in L_p and *g* does not change the phase difference between the two elements, as the difference in phase depends on the orientation of

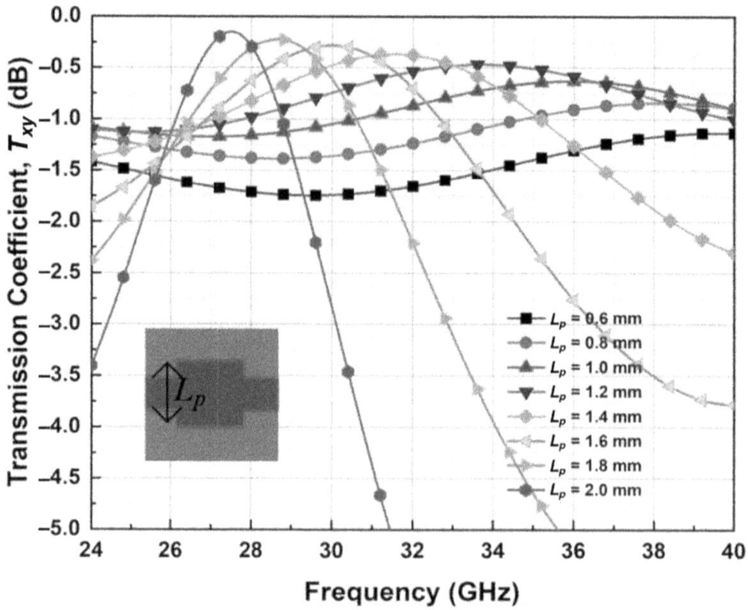

Figure 8.15 Changes in transmission coefficient magnitude as a function of Lp.

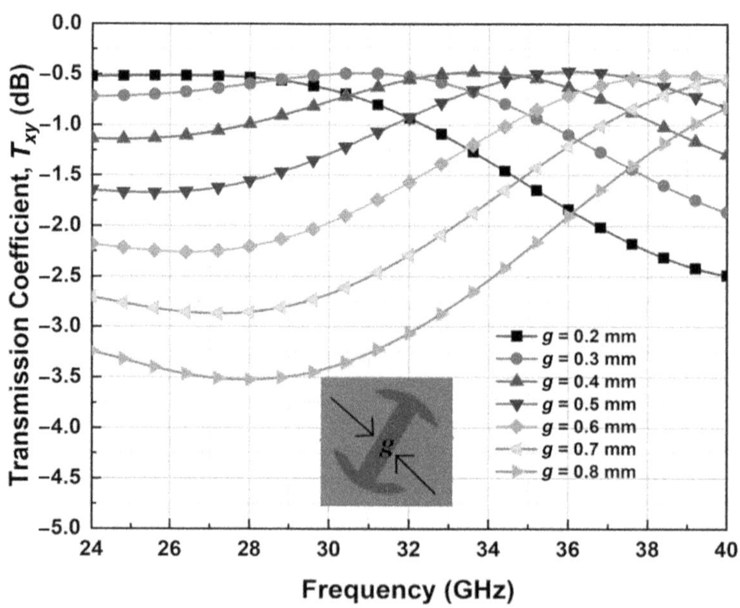

Figure 8.16 Changes in the magnitude of the transmission coefficient as a function of 'g'.

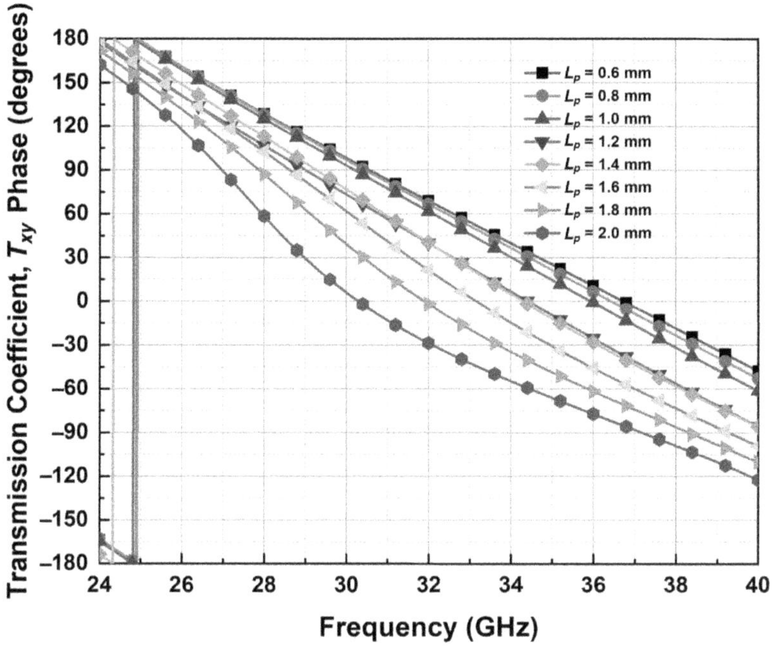

Figure 8.17 Simulated transmission coefficient phase of the phase shift element with variations in L_p (bit-0).

the dipole element. The change in phase of the UC elements with changing L_p is depicted in Figures 8.17 and 8.18. These figures demonstrate that although the phase of each element changes with variations in L_p, the difference in phase of elements is unchanged at 180 degrees. Likewise, a simulation investigation is undertaken to analyze how changes in the 'g' affect the difference in the phase of a UC. As depicted in Figures 8.19 and 8.20, it can be summarized that the alteration of 'g' has a minimal impact on the phase difference. The suggested elements demonstrate a transmission loss of about 1.5 dB while maintaining a consistent phase difference of 180 degrees between the elements across the 24–40 GHz frequency range.

8.3.2 Feed antenna design and simulation

A printed dipole antenna with four directors designed to operate within the 26.3–29.3 GHz range forms the basis for a microstrip Yagi antenna. Three antennas, each consisting of three individual antenna elements as depicted in Figure 8.21, are organized in a nine-element configuration to feed the transmitarray panel, with a spacing of 10 mm.

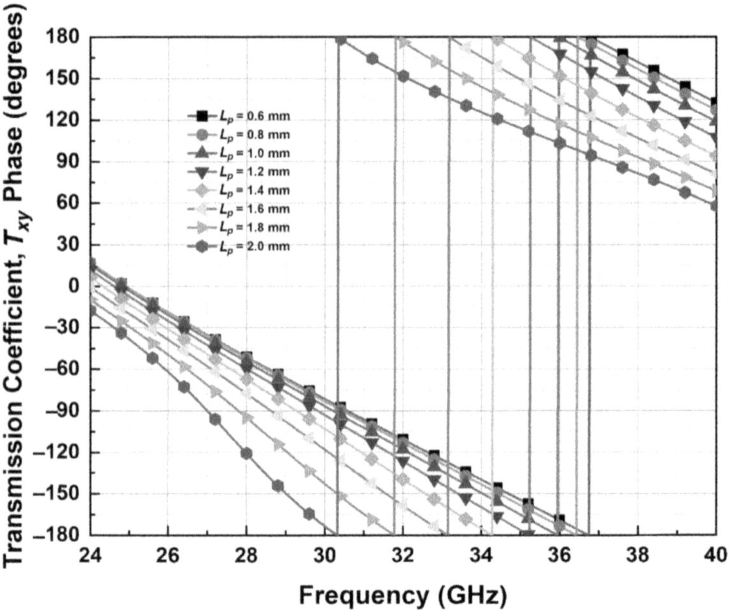

Figure 8.18 Simulated phase of the transmission coefficient with varying L_p (bit-1).

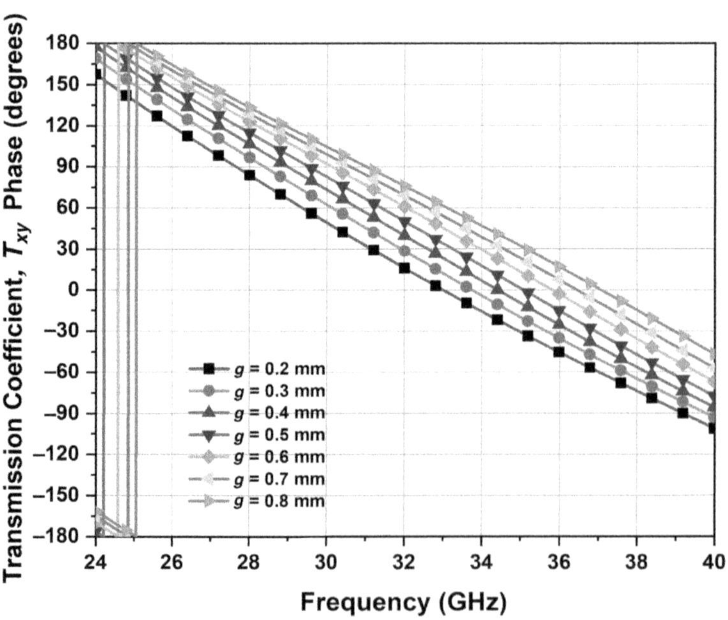

Figure 8.19 Simulated phase of the transmission coefficient with varying g (bit-0).

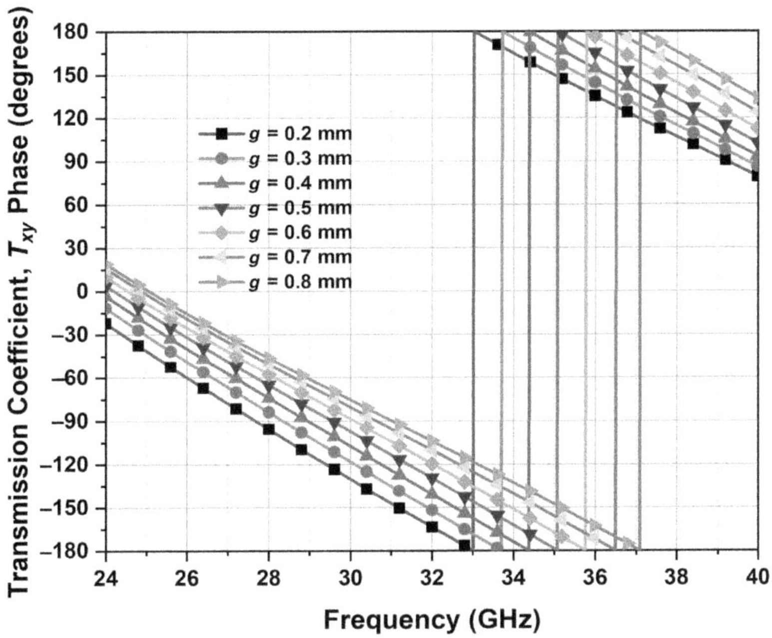

Figure 8.20 Simulated phase of the transmission coefficient with varying g (bit-1).

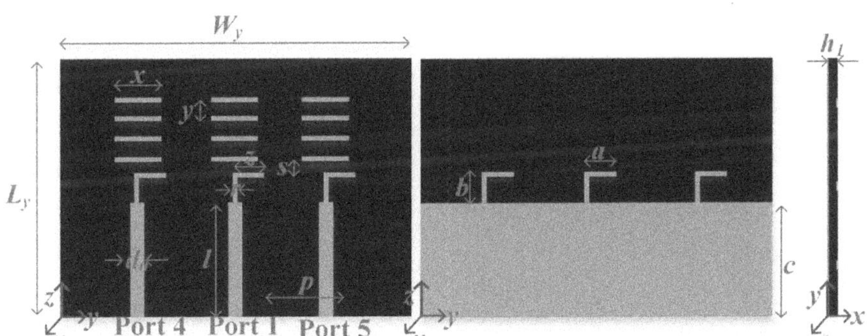

Figure 8.21 Three-port Yagi antenna employed for feeding the 2-D beam-switching antenna. (a and b) Side views, (c) top view with dimensions.

Figure 8.22 displays the scattering parameters of the feed Yagi antenna, functioning within the 26.3–29.3 GHz range. Furthermore, the array elements demonstrate a mutual coupling of >20 dB.

8.3.2.1 Transmitarray design

Organizing the elements based on the distribution depicted in Figure 8.23 gives the design of a transmitarray, which is powered by a 9-port Yagi feed array, resulting in the creation of a 2-D switchable beam antenna. Figures 8.24 and 8.25 show the perspective visualization of the 2-D beamforming antenna and the simulated S-parameter results of the transmitarray. Figure 8.26 depicts the simulated 3-D patterns of the transmitarray at 28 GHz for port 1, port 5, port 3, and port 7.

When designing the transmitarray, a crucial factor to consider is its aperture efficiency (AE). The AE of the antenna is determined using the equation referenced in [17]. It primarily depends on both the aperture area and the gain of the antenna. At 28 GHz, the (AE) for the proposed transmitarray is determined to be 23.6%.

8.3.2.2 Fabrication and measurement

The fabricated design of the 2-D switchable beam antenna is outlined in Figure 8.27. Maintaining accurate dimensions during fabrication is essential for antenna systems operating at millimeter-wave frequencies, as minor deviations may result in undesirable responses. Hence, careful consideration is given to accurately print the dimensions on the substrate to

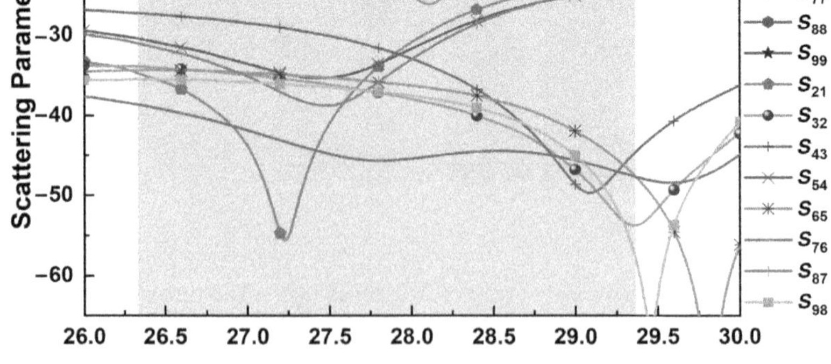

Figure 8.22 Simulated S-parameter results for the 9-port Yagi antenna employed as the feeding array.

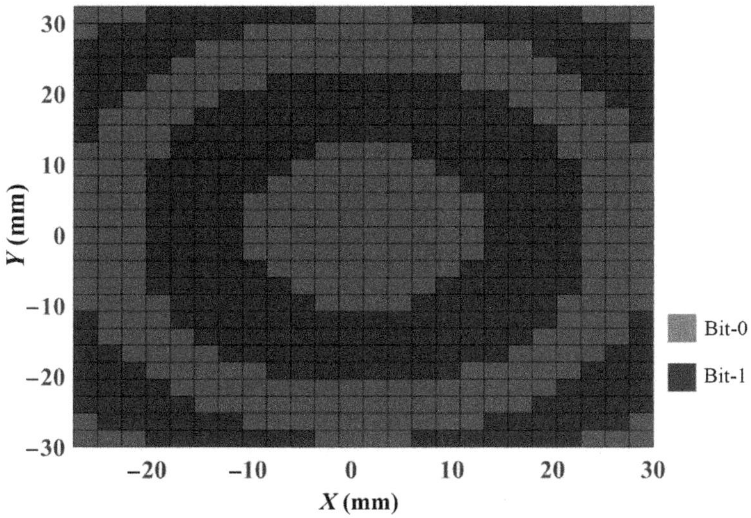

Figure 8.23 Discrete phase distribution of the 2-D switchable beam transmitarray at a focal distance of f = 30 mm and frequency of 28 GHz.

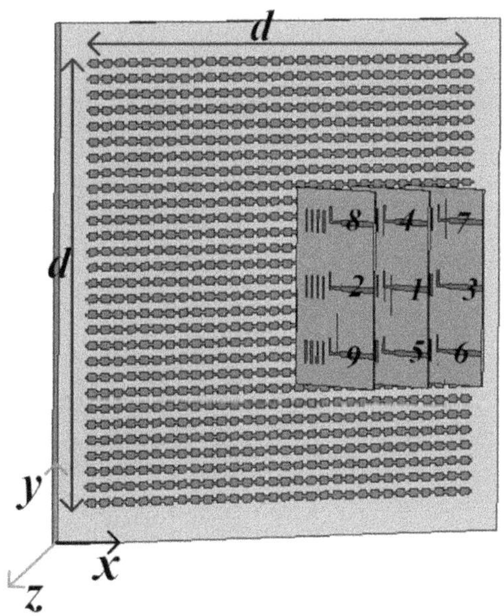

Figure 8.24 Perspective visualization of the 2-D beamforming antenna.

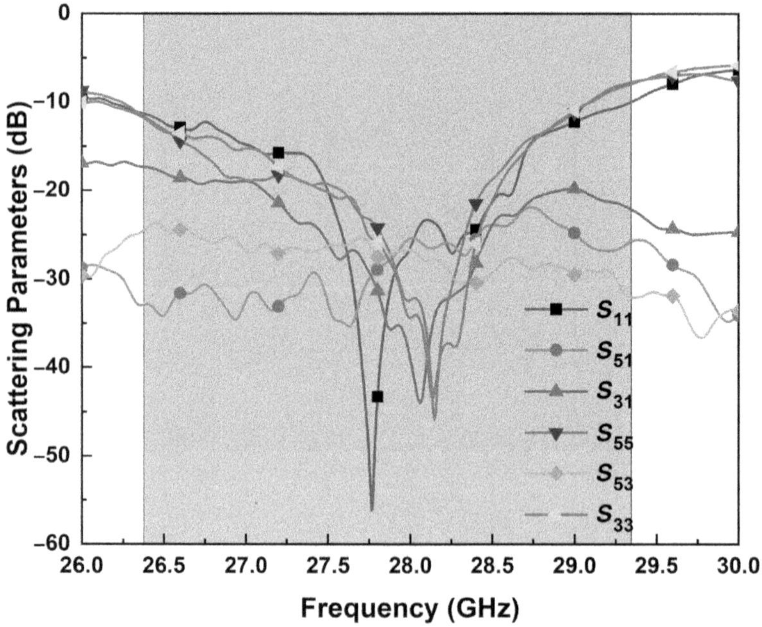

Figure 8.25 Simulated *S*-parameter results of the transmitarray.

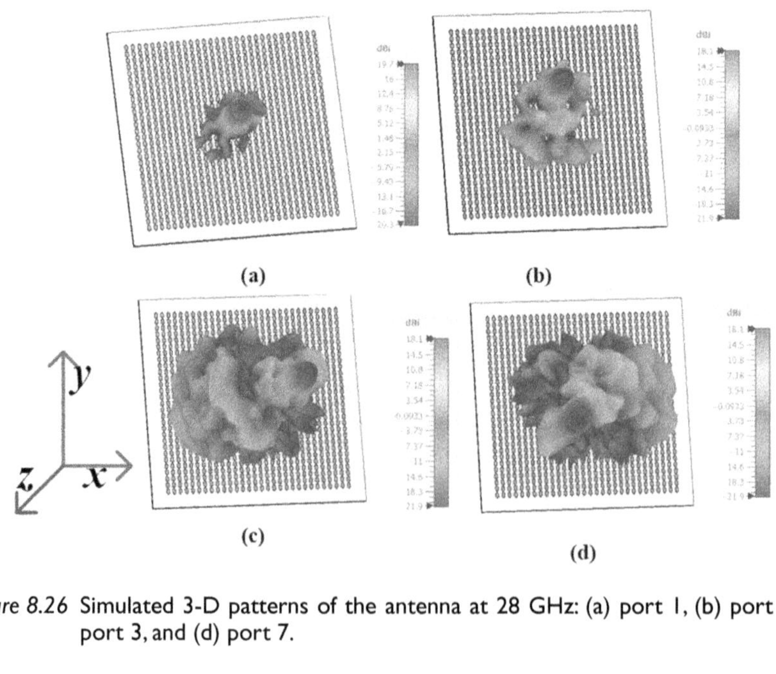

Figure 8.26 Simulated 3-D patterns of the antenna at 28 GHz: (a) port 1, (b) port 5, (c) port 3, and (d) port 7.

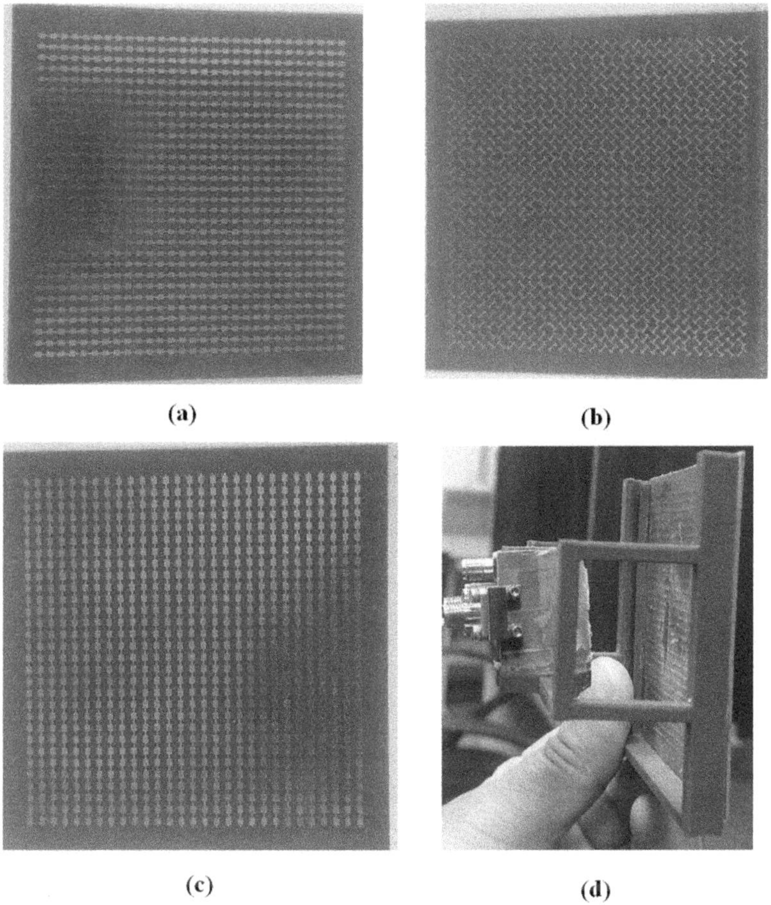

Figure 8.27 Prototype of the proposed transmitarray.

prevent such discrepancies. Each element is fed using a Hirose connector (HK-LR-SR2(12)).

The performance parameters undergo experimental validation in the anechoic chamber, utilizing a vector network analyzer (VNA) (N5224B). Figure 8.28 illustrates the measured results of the proposed antenna, indicating an operational instantaneous maximum percent width (IMPW) of 10.8% (26.3–29.3 GHz) with isolation exceeding 20 dB.

The 2-D patterns of the switchable beam antenna are measured across different frequencies in the *xz*- and *yz*-planes, as depicted in Figures 8.29–8.32. The patterns' results exhibit an HPBW spanning from 9.8 to 11.2 degrees across different switchable patterns.

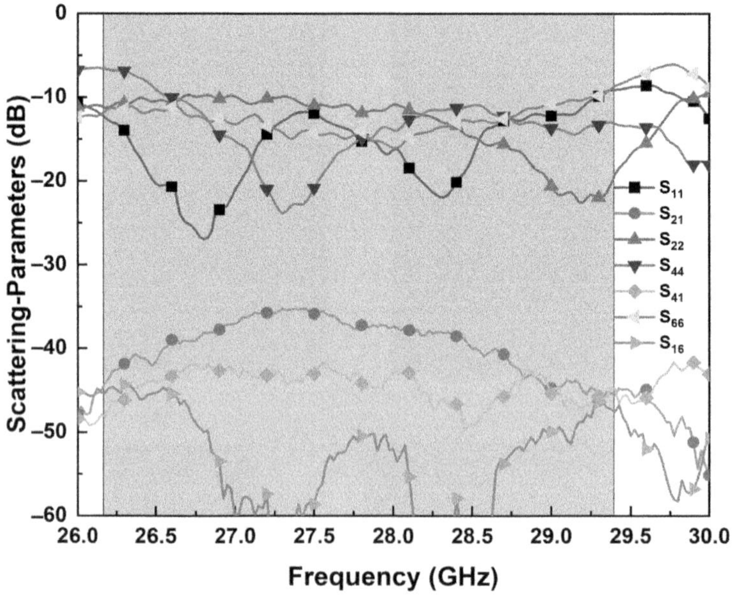

Figure 8.28 Measured *S*-parameters of the transmitarray.

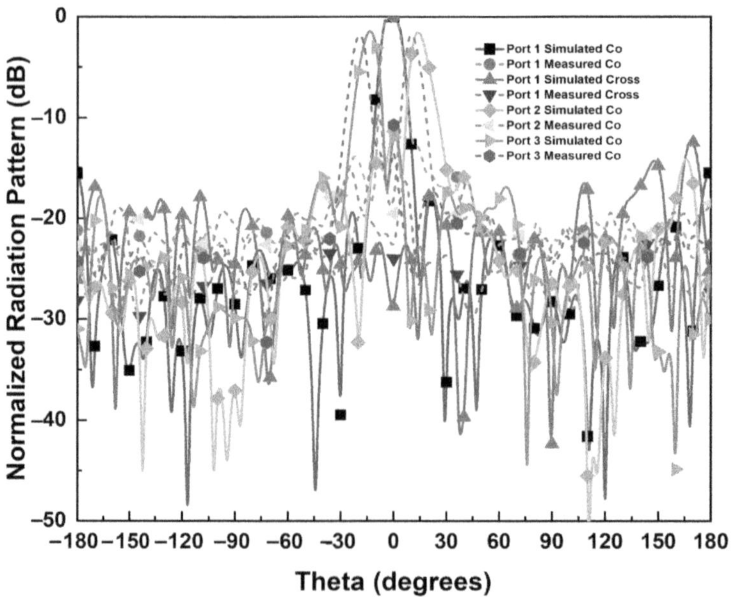

Figure 8.29 Simulated and measured radiation patterns at 27 GHz in *xz* plane.

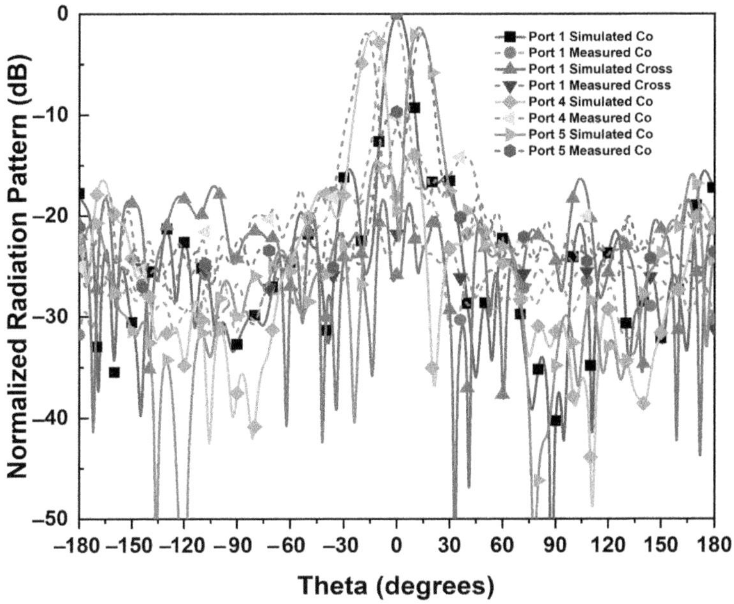

Figure 8.30 Simulated and measured radiation patterns of the transmitarray at 27 GHz in *yz*-plane.

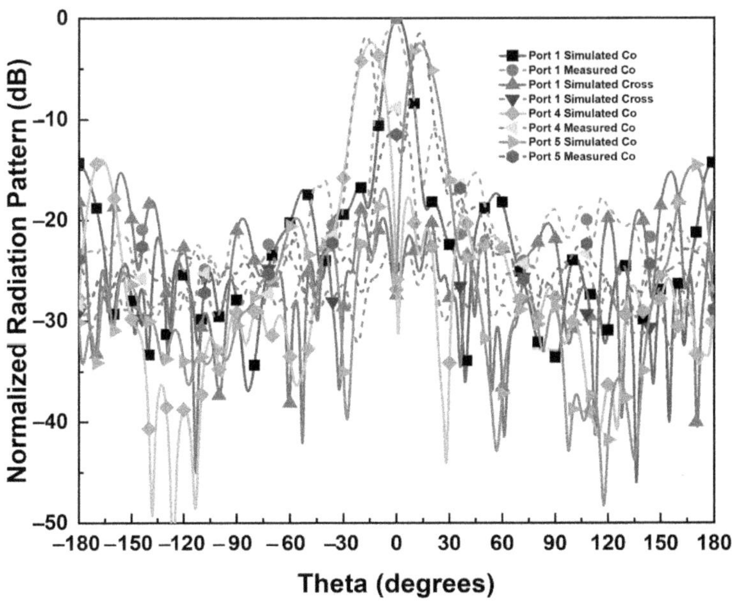

Figure 8.31 Simulated and measured radiation patterns of the transmitarray at 29 GHz in *xz* plane.

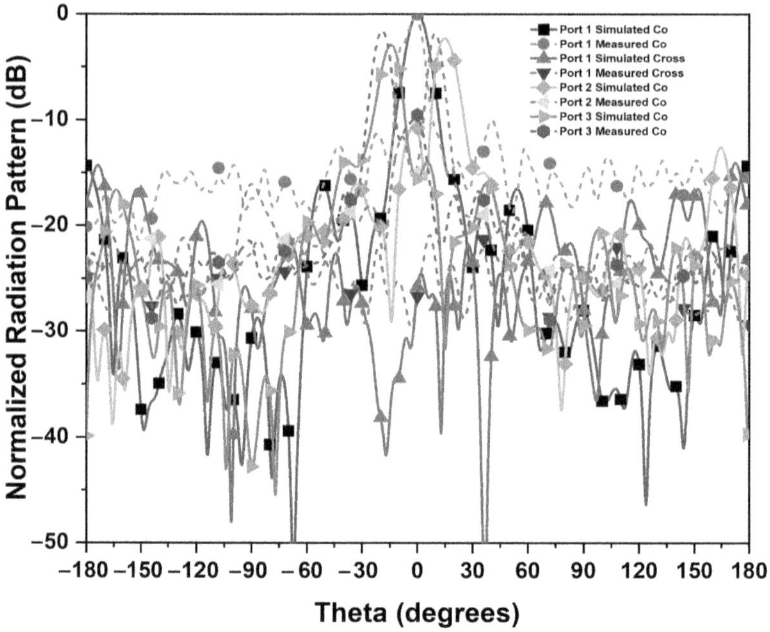

Figure 8.32 Simulated and measured radiation patterns transmitarray at 29 GHz in *yz* plane.

Figure 8.33 illustrates the gain associated with various ports of the antenna. Port 1 attains a measured peak gain of around 19.7 dBi, while a steering loss of 1.5 dB is noted in the remaining ports.

8.4 CONCLUSION

This chapter presents a compact transmitarray antenna designed for millimeter-wave applications, capable of 1-D and 2-D beam switching. The panel utilizes a 3-port patch MIMO array and a 9-port antenna array as a feed mechanism to enable beam switching. The proposed designs offer practical applications as repeater antennas, especially in public transport vehicles to direct beams toward occupants. Furthermore, the antenna shows potential for millimeter-wave base stations to focus electromagnetic waves on specific users.

Its scalability enables the extension of the transmitarray panel to incorporate additional antennas, rendering it a feasible choice for implementing massive MIMO antenna systems.

To achieve wideband operation with a single beam, a K_a-band horn antenna proves effective in feeding the transmitarray panel. Additionally,

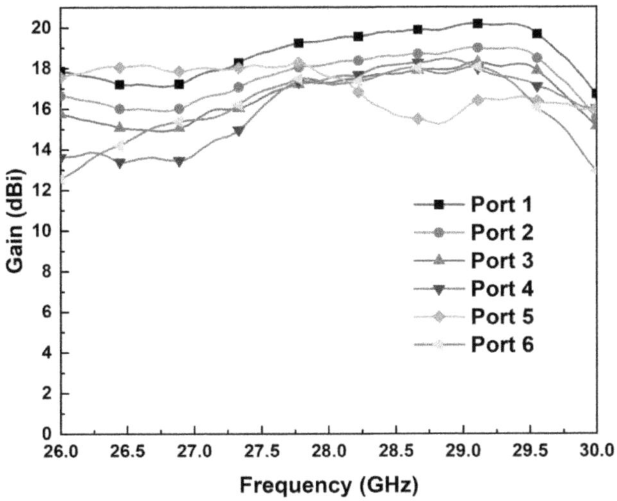

Figure 8.33 The measured gain of the transmitarray for various antenna ports in the direction of maximum radiation.

augmenting the phase levels within the panel could result in higher gain and improved AE. Furthermore, integrating a 1:3 and 1:9 microwave switch with the feed antenna facilitates efficient implementation of beam switching in real-world applications. By incorporating a programmable microcontroller unit alongside the 1:3 and 1:9 microwave switch, different beam directions can be programmed as required, thus providing flexibility and adaptability.

REFERENCES

[1] M. Alibakhshikenari et al., "A comprehensive survey on antennas on-chip based on metamaterial, metasurface, and substrate integrated waveguide principles for millimetre-waves and terahertz integrated circuits and systems," IEEE Access, vol. 10, pp. 3668–3692, 2022, doi: 10.1109/ACCESS.2021.3140156.

[2] T. S. Rappaport et al., "Millimetre wave mobile communications for 5G cellular: it will work!," IEEE Access, vol. 1, pp. 335–349, 2013, doi: 10.1109/ACCESS.2013.226081.

[3] W. Hong et al., "Multibeam antenna technologies for 5G wireless communications," IEEE Transactions on Antennas and Propagation vol. 65, no. 12, pp.6231–6249, 2017, doi: 10.1109/TAP.2017.2712819.

[4] R. Mailloux, Phased Array Antenna Handbook, Third Edition, Artech, 2017.

[5] B. He, Y. Jiao, and H. He, "Design of high-gain lens antenna for 5G application," 2017 Sixth Asia-Pacific Conference on Antennas and Propagation (APCAP), 2017, pp. 1–3, doi: 10.1109/APCAP.2017.8420863.

[6] A. H. Abdelrahman, F. Yang, A. Z. Elsherbeni, P. Nayeri, ans C. A. Balanis, Analysis and Design of Transmitarray Antennas, Morgan & Claypool, 2017.
[7] X. Yang, Y. Zhou, L. Xing, and Y. Zhao. A wideband and low profile transmitarray antenna using different types of unit-cells. Microwave and Optical Technology Letters, vol. 61, no. 6, pp. 1584–1589, 2019. doi: 10.1002/mop.31754.
[8] M. -Y. Li, Y. -L. Ban, and F. -Q. Yan, "Wideband low-profile Ku-band transmitarray antenna," IEEE Access, vol. 9, pp. 6683–6688, 2021, doi: 10.1109/ACCESS.2020.3048346.
[9] X. Yi, T. Su, B. Wu, J. Chen, L. Yang, and X. Li, "A double-layer highly efficient and wideband transmitarray antenna," IEEE Access, vol. 7, pp. 23285–23290, 2019, doi: 10.1109/ACCESS.2019.2893608.
[10] C. Jouanlanne et al., "Wideband linearly polarized transmitarray antenna for 60 GHz backhauling," IEEE Transactions on Antennas and Propagation, vol. 65, no. 3, pp. 1440–1445, Mar. 2017.
[11] A. Moknache et al., "A switched-beam linearly-polarized transmitarray antenna for V-band backhaul applications," in Proceedings of the 10th European Conference on Antennas and Propagation, pp. 1–5, 2011.
[12] Y. Cai et al., "A novel ultrawideband transmitarray design using tightly coupled dipole elements," IEEE Transactions on Antennas and Propagation, vol. 67, no. 1, pp. 242–250, Jan. 2019, doi: 10.1109/TAP.2018.2878079.
[13] P. Mei, G. F. Pedersen, and S. Zhang, "A broadband and FSS-based transmitarray antenna for 5G millimeter-wave applications," IEEE Antennas and Wireless Propagation Letters, vol. 20, no. 1, pp. 103–107, Jan. 2021, doi: 10.1109/LAWP.2020.3042295.
[14] K. Mavrakakis, H. Luyen, J. H. Booske, and N. Behdad, "Wideband transmitarrays based on polarization—rotating miniaturized-element frequency selective surfaces," IEEE Transactions on Antennas and Propagation, vol. 68, no. 3, pp. 2128–2137, March 2020, doi: 10.1109/TAP.2019.2949694.
[15] P. Feng, S. Qu, and S. Yang, "Octave bandwidth transmitarrays with a flat gain," IEEE Transactions on Antennas and Propagation, vol. 66, no. 10, pp. 5231–5238, Oct. 2018.
[16] Y. Ge, C. Lin, and Y. Liu, "Broadband folded transmitarray antenna based on an ultrathin transmission polarizer," IEEE Transactions on Antennas and Propagation,vol. 66, no. 11, pp. 5974–5981, Aug. 2018.
[17] Z. Wani, M. P. Abegaonkar, and S. K. Koul, "Thin planar metasurface lens for millimeter wave MIMO applications," IEEE Transactions on Antennas and Propagation, vol. 70, no. 1, pp. 692–696, doi: 10.1109/TAP.2021.3098571.
[18] G. Li, Z. Wang, and Y. Ge, "Multibeam folded transmitarray antenna for massive MIMO applications," in 2019 International Workshop on Electromagnetics: Applications and Student Innovation Competition (iWEM), pp. 1–2, 2019, doi: 10.1109/iWEM.2019.8887895.

Chapter 9

An array of highly directive printed fractal radiators for D2H applications

Kalyan Sundar Kola and Anirban Chatterjee

9.1 INTRODUCTION

Modern communication systems rely heavily on microstrip radiators for a variety of purposes [1,2]. Patch antennas are the best choice as they are small, portable, sturdy, inexpensive, and easy to set up [1,2]. Due to these characteristics, they are well-suited for an extensive range of applications, such as electronic support measures, mobile communication devices, and radar systems [1,2]. Data transmission, satellite telephone, and television are among the several applications that use the Ku-band spectrum in the early morning hours. Microstrip radiators are widely used because they can function in the ITUs Ku-band frequency range of 12.20 to 12.70 GHz. Linearly polarized antennas considerably impact satellite applications [1–4]. The antennas feature improved signal reception, lower signal degradation, stronger multipath interference resistance, flexible deployment options, satellite system compatibility, and decreased interference sensitivity. Direct broadcast satellite (DBS) applications are well-suited for linearly polarized antennas due to their dependability and favorable signal-to-noise ratio. Nevertheless, conventional patch antennas are constrained in their maximal bandwidth by a quality factor (Q_n) [13–21].

Murata *et al.* [6] demonstrated the practical implementation of a planar antenna for DBS service and other Ku-band applications. A printed antenna with high gain, two ports, and two polarizations was shown in a study by Chen *et al.* [7].To efficiently and affordably receive mobile DBS broadcasts, Mousavi *et al.* [8] developed a phased array technology. Microstrip patch antennas with slots in their architectural design provide better gain and bandwidth, according to the studies listed in [9–11]. The initial proposition of the hybrid split quadrilateral antenna was attributed to Alam *et al.* [12]. A wide range of frequencies is efficiently used by the antenna design [11]. A tiny antenna in the form of an epsilon was suggested by Jaiverdhan *et al.* [13]. This antenna's ultra-wideband property allows it to function in two separate frequency bands. In their study, Kola *et al.* introduced a range of printed fractal antennas designed for various wireless applications

DOI: 10.1201/9781003470854-9

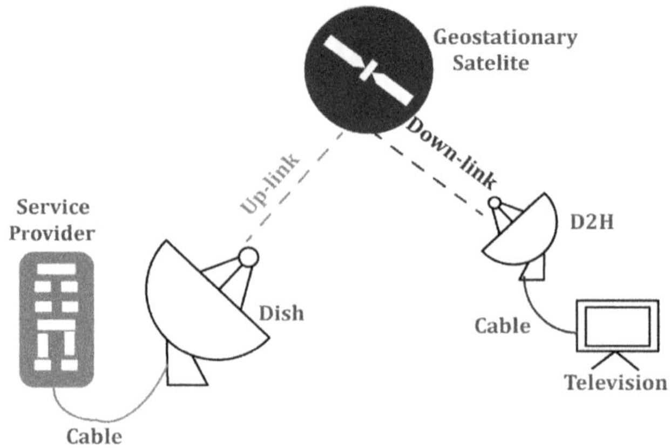

Figure 9.1 Architectural representation of direct-to-home (D2H) services.

[14–22]. For use with Ku-band satellites, a number of printed antennas have been developed. In addition to a high radiation efficiency, minimal cross-polarization, and negligible Electromagnetic Interference electromagnetic compatibility (EMC) effects, these antennas provide substantial gain [23–24].

Figure 9.1 of this chapter shows an architectural depiction of direct-to-home (D2H) services. To address the needs of D2H applications in the Ku-band, this chapter primarily suggests a printed radiator based on Minkowski fractals and then a two-element linear array.

Some of the novel aspects of the designed antennas are outlined below:

- The sole radiator is derived from Minkowski's second iteration.
- The sole antenna is capable of offering 0.20 GHz impedance bandwidth.
- It also offers below –50dBi of cross-polarization suppression along its main beam direction.
- It also offers a very high absolute gain of 10.20 dBi at a particular frequency.
- To minimize losses, maximize isolation, and enhance bandwidth responsiveness, the feed network for the array was developed using the Wilkinson power divider [21].
- At the optimal radiation point, the array offers decent bandwidth, impressive gain, and shallow x-pol. suppression.

In addition to the benefits already stated, the array and the sole radiator have a radiation efficiency of over 88% and more than 67% of aperture efficiency [27,29], respectively. Both antennas exhibit moderate electromagnetic interference effects at suitable resonant frequencies because they provide correction factors of more than 30 dB/m [30]. Considering their specs, both antennas are great options for D2H Ku-band networks.

9.2 THE SOLE RADIATOR

The subsequent sections describe the formation of the proposed fractal geometry in great detail, compute the contour area, provide a comprehensive embodiment, calculate the circuit equivalent of the solitary radiator, and analyze the simulated results for each.

9.2.1 Development of the fractal structure

Figure 9.2 illustrates the progressive development of the fractal structure under consideration. At the outset, a geometric square 'A' with a side length of 'a' has been selected, as illustrated in Figure 9.2(a). The four sides of 'A' have been modified to include the Minkowski first iteration method, and the Minkowski squares have been indicated on 'A'. As shown in Figure 9.2(b), the Minkowski square-based geometry 'B' is implemented. In addition, by etching those squares, the fractal structure 'C'as illustrated in Figure 9.2(c) is produced. 'a_1' is the side length of the Minkowski squares. Following this, the structure 'C' is reduced in size by one-third, resulting in the square 'D' that is illustrated in Figure 9.2(d). 'a_2' indicates the side length of the structure 'D'. Moreover, by subtracting and adding the structures 'C' and 'D', the newly devised fractal geometry 'E' and 'F' are produced; these are denoted in Figure 9.2(e) and (f), correspondingly. The placement of structure 'E' at the four corners of the *xy*-plane ensures that the antenna will

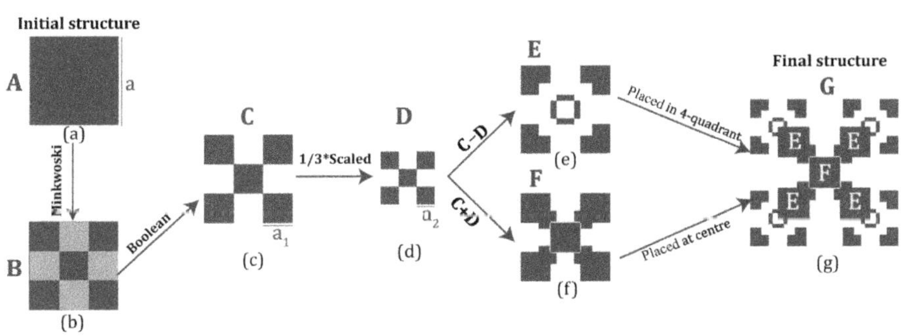

Figure 9.2 Creating the suggested complex shape one step at a time. (a) initial structure, (b) Minkowski structure based one quadrant, (c) slotted one quadrant, (d) scaled version of C, (e) Boolean structures of C & D, and (f) four-folded geometry.

Table 9.1 Dimension details of the intended antennas (unit: mm)

Parameter	*Value*	*Parameter*	*Value*	*Parameter*	*Value*	*Parameter*	*Value*
a	9.30	a_1	3.10	a_2	1.03	W	49.60
S_L	8.54	S_W	1.25	L_F	27	W_A	94.95
L_A	64.42	h_s	0.787	h_t	0.017	R	1

produce a symmetric radiation pattern. Structure 'F' is positioned in the exact middle of the *xy*-plane for better parametric results. Ultimately, the fractal geometry 'G' is attained by converting both structures to Boolean values, as illustrated in Figure 9.2(g). The numerical values of the dimension notation for each of the subcomponents are detailed in Table 9.1.

9.2.2 Contour area computation

The proposed fractal structure 'G' consists of a few intermediate structures like A,C,D,E,F. The effective contour areas of the respective structures are computed [22] as follows:

$$A_A = a^2 \tag{9.1}$$

$$A_C = A_A - (4\times a^2) = \left(\frac{5}{9}\times a^2\right) \tag{9.2}$$

$$A_D = 5\times a_2{}^2 = \left(\frac{5}{81}\times a^2\right) \tag{9.3}$$

$$A_E = A_C - A_D = \left(\frac{40}{81}\times a^2\right) \tag{9.4}$$

$$A_F = A_C + A_D = \left(\frac{50}{81}\times a^2\right) \tag{9.5}$$

$$A_G = (4\times A_E) + A_F = \left(\frac{210}{81}\times a^2\right) \tag{9.6}$$

After putting all the parameter values in Equations (9.1) to (9.6), the computed [22] effective contour area of the intended geometry becomes 2.24 cm^2.

9.2.3 Complete embodiment

The low-cost Rogers 5880 material (ε_r =2.2) with a PCB laminated surface is covered with the suggested fractal structure 'G'. As a complete base, the opposite layer of material is used. The ground plane construction is not flawed in any way, which means the resonating frequency and gain are both kept constant. The primary beam may be directed at 0° by including three stubs ($S_L \times S_W$) at the center of structure 'G', with each stub forming a right angle to the other. Figure 9.3 illustrates the geometry of the single-layered antenna. A frontal view of the complete embodiment is illustrated in Figure 9.3(a), whereas a rear view is depicted in Figure 9.3(b). The thickness of the ground and radiating layers is 0.017 mm, while the substrate layer has a thickness of 0.787 mm. Table 9.1 contains the exact measurements of a single radiator. The radiator was designed and tested using the SONNET EM simulator, with CST Microwave Studio (v2019) serving as a cross-verifier.

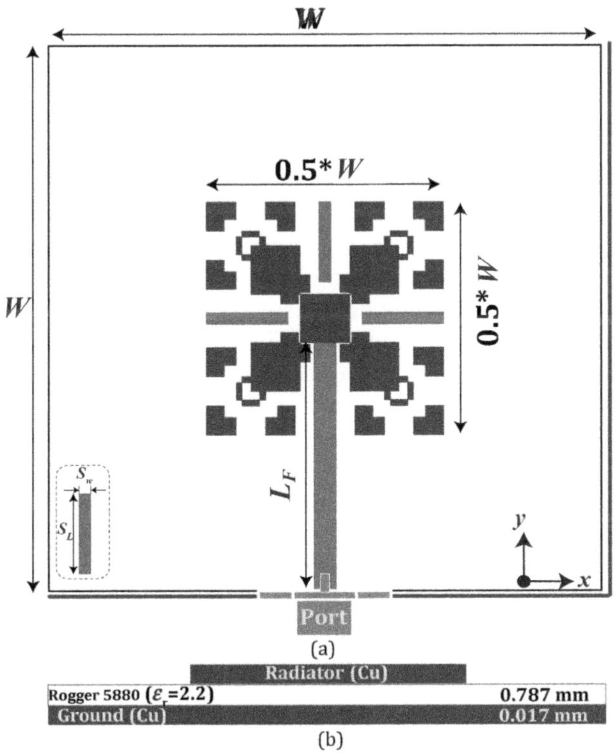

Figure 9.3 Proposed printed antenna.

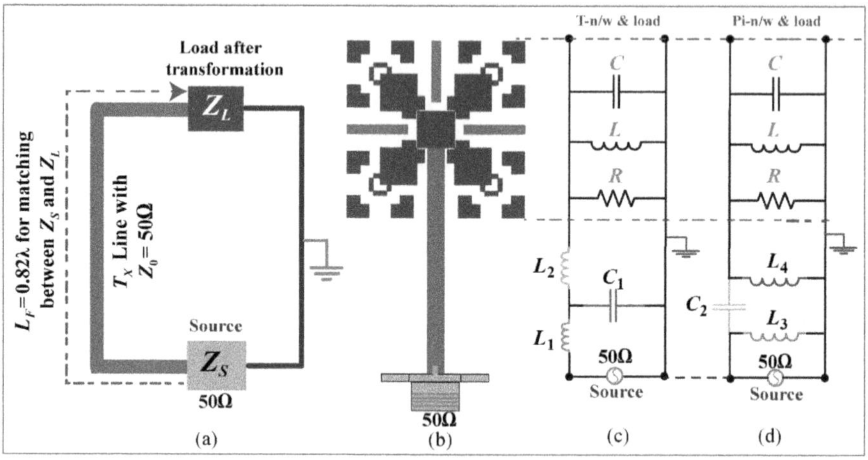

Figure 9.4 Antenna circuit model: (a) rudimentary model; (b) isolated radiator; (c) aggregated load model with T; and (d) Pi-networks.

9.2.4 Circuit model of the antenna

An analysis of the antenna under discussion has been conducted by the use of circuit equivalent techniques, namely the principles of transmission line theory. The matching network for the feedline consists of T and Pi types, whereas the parallel RLC circuit represents the antenna's radiator. Figure 9.4(a) shows a circuit model with a single radiator. In the current setup, the source and load are impedance-matched using a transmission line that is 0.82 times the wavelength. Figure 9.4(b) shows the antenna, which has an impedance of 50Ω at its source port. Figure 9.4(c) illustrates the radiator of the antenna, whereas Figure 9.4(d) depicts the matching networks for the T and Pi varieties. Circuit model scattering parameter $|S_{11}|$ and its relevance to the nodal-quality factor (Q_n), as shown in Figure 9.5(a) [25]. Figure 9.4(a) shows that the proposed antenna design uses a parallel RLC load. The performance of the T/Pi network is evaluated in Figure 9.5(b) by looking at the value of the $|S_{11}|$ parameter. An exact reproduction of the circuit design was achieved using the Cadence simulator. The model attains optimal resonance at a frequency of 12.50 GHz, and it has a favorable impedance bandwidth response, according to empirical data. The impedance characteristics of the circuit model coupled to the antenna are illustrated in Figure 9.5(c). Under magnification, Figure 9.5(c) demonstrates that the two antennas offer an exact resistive match of 50Ω. The computed values of the lumped elements are listed in Table 9.2.

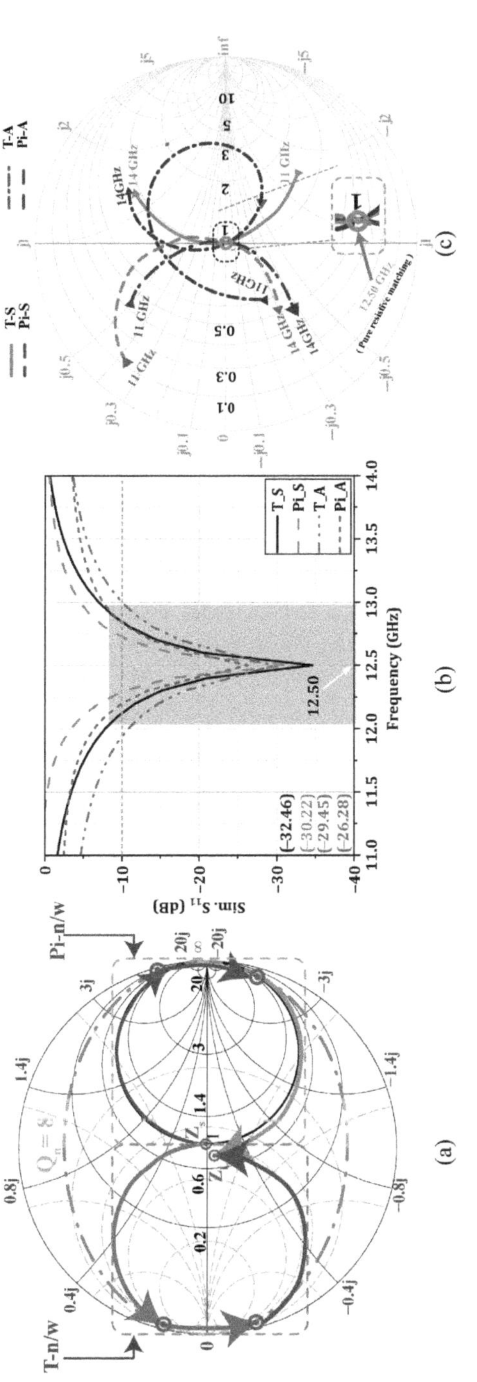

Figure 9.5 Antenna and array circuit model showed: (a) the Q_n circle; (b) the synthetic S_{11}; and (c) The networks matched with their resistance.

9.3 PROPOSED ARRAY

A linearly polarized array was developed for the intended uses to enhance parametric outcomes. The equivalent circuit model followed by the construction of the array has been carried out in this section, respectively.

9.3.1 Equivalent circuit model

The circuit architecture of the array is illustrated in Figure 9.6(a). Wilkinson power divider feed network supplies array input. Feedlines, which symbolize the T and Pi sub-blocks within a power divider, are employed to optimize the device's design. The corresponding sub-block lumped equivalent circuit models are illustrated in Figure 9.6(b). Figure 9.5(b) shows the magnitude of the $|S_{11}|$ parameter, whereas Figure 9.5(c) shows the impedance responses of the array model. According to the results, the resonance frequency is 12.50 GHz. With the array set up as planned, the impedance bandwidth is now higher. Existing circuit models ignore the effects of mutual coupling, dielectric materials, and fringing fields. Thus, these considerations may account for the broad frequency range seen in practical settings. When the model yielded positive results, the SONNET simulator was used to produce the array [23]. To guarantee the correctness and trustworthiness of the

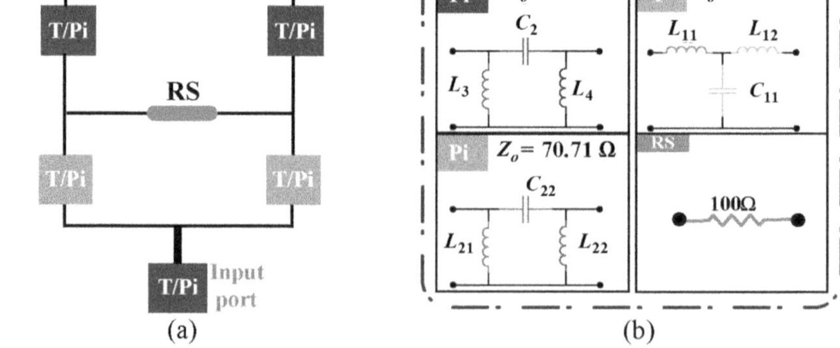

Figure 9.6 Array circuit model: (a) array with feed network, (b) lumped model of each section.

Table 9.2 Lumped element values of the circuit model of antenna and array

Type		*Element*	*Value*	*Element*	*Value*	*Element*	*Value*
Matching n/w	Load	R	49.82 Ω	L	0.63nH	C	0.28pF
50 Ω	T-n/w	L_1	3.52nH	L_2	3.52nH	C_1	0.08pF
	Pi-n/w	L_3	0.07nH	L_4	0.07nH	C_2	1.56nH
70.7 Ω	T-n/w	L_{11}	0.84nH	L_{12}	0.84nH	C_{11}	0.17pF
	Pi-n/w	L_{21}	0.45nH	L_{22}	0.45nH	C_{22}	0.24pF

findings, the array was verified using CST Microwave Studio (v2019) using a cross-validation approach [24]. All the lumped element values are listed in Table 9.2.

The input impedance (Z_{in}) can be ascertained as follows:

$$Z_{in} = Z_0 \frac{Z_L + jZ_0 tan\beta l}{Z_0 + jZ_L tan\beta l} \tag{9.7}$$

where Z_L and Z_0 are indicated as antenna's load and characteristics impedance, respectively, whereas βl is its length.

9.3.2 Construction of array geometry

By using a handful of antennas, a linear array may be constructed to get enhanced parametric results. The proposed linear array is depicted in Figure 9.7. It can be clearly observed that the antenna elements are placed side by side inside the array geometry. This setup aims to achieve an acceptable radiation pattern in a proper direction. To mitigate the antenna's mutual coupling effect and also boost antenna gain, a fixed inter-element spacing of 0.5λis fixed. The feed topology of the array has been configured by a distinctive Wilkinson power divider [21]. All 50Ω and 70.7Ω transmission line lengths have been fixed as the integer multiple of 0.5λ and 0.25λ, respectively. Quarter-wavelength isolation from the feed network is improved by mounting a resistor R in the junction, as shown in Figure 9.7. Table 9.1 provides the quantitative figures for the array's dimension notation.

9.4 RESULTS AND DISCUSSION

The proposed Minkowski fractal-based printed radiator, followed by a linear array, has been designed for D2H applications. Models of both antennas were created in SONNET and then cross-checked in CST microwave studio (v2019). The antenna is designed with the full ground plane in

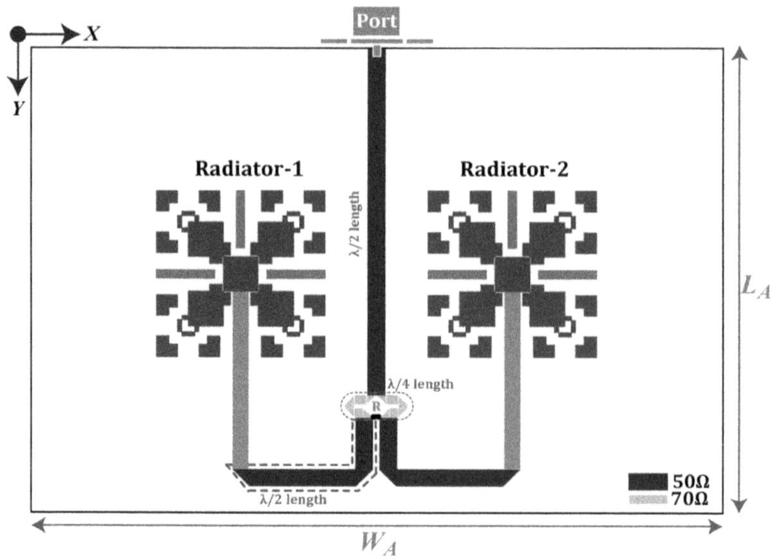

Figure 9.7 Proposed two-element linear array.

consideration. There is no flawed construction included, which might cause the resonant frequency to wander and the gain to decrease. This section will cover the discussion about the obtained simulated results of the intended antennas.

Figure 9.8(a) shows that the sole radiator, which is resonated at 12.50 GHz, has a return loss of 36.4 dB. Antenna characteristics at the designated frequency include an impedance bandwidth of 0.20 GHz. The absolute gain characteristics of the antenna are illustrated in Figure 9.8(b) and it presents the antenna gain of 10.20 dBi. The field patterns (*E*, *H*) of the single radiator in Figure 9.8(c) offer a cross-pol. discrimination of –57dBi at cut-angle 0°, which is sufficient and required in this context. Figure 9.8(d) illustrates the *xz*- and *yz*-radiation patterns of the antenna. Conspicuously, the backlobe radiation is less than –25 dBi and the primary beam of the antenna is oriented at 0°. The antenna has effectively achieved a symmetric radiation pattern, as illustrated in Figure 9.8(e), through the utilization of its three-dimensional radiation pattern. The numerical values pertaining to the antenna parameters are presented in Table 9.3.

The aperture efficiency [25, 26] of an antenna is calculated as follows:

$$\eta_{ap} = \frac{D}{D_{max}} \tag{9.8}$$

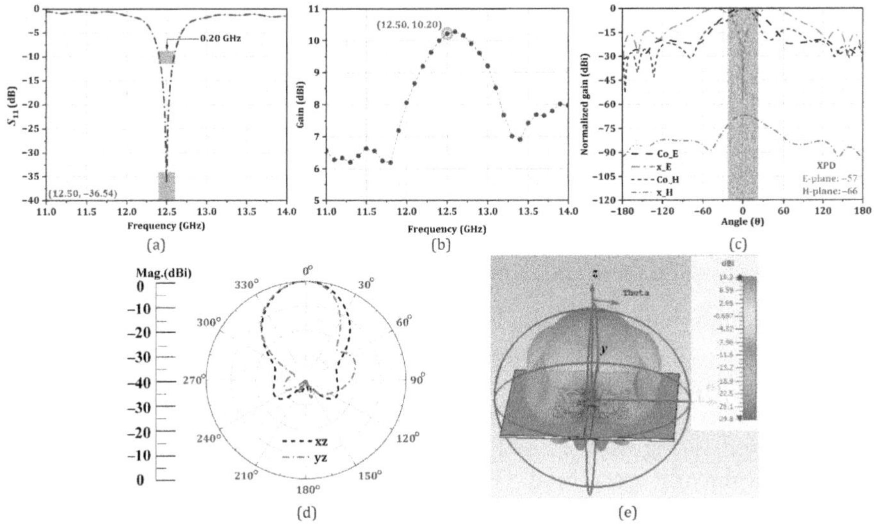

Figure 9.8 Single antenna's outcomes in terms of return-loss, gain, patterns (*E*,*H*,*xz–yz*,3D).

Table 9.3 Parametric performances of the antenna and the array

Parameter(s)	*Single antenna*	*Antenna array*
f_r(GHz)	12.50	12.44
S_{11}(dB)	−36.54	−20.42
BW (GHz)	0.20	0.26
Gain (dBi)	10.20	14.20
Directivity (dBi)	11.34	15.67
XPD (dBi)	-57	-42
Rad. eff. (%)	91	89
App. eff. (%)	72	68
CF (dB/m)	33.24	32.16

$$D_{max} = \frac{4 \times \Pi \times A}{\lambda^2} \quad (9.9)$$

where D and A represent the antenna's directivity and total area, respectively.

The isolated antenna possesses a directivity of 11.34 dBi and an overall surface area of $49.6 \times 49.6\text{mm}^2$. This results in an increase in the predicted aperture efficacy of the antenna to 72% at 12.50 GHz. Additionally, 91% radiation efficiency is achieved by the single radiator [27–29]. It is evident that the single antenna performs very well for D2H applications in the Ku-band.

One of the most essential aspects of an antenna's electromagnetic impact is its EMC. Using the surface equivalence technique, the antenna's correction factor (CF) [30] is determined, and the CF is then calculated as follows:

$$CF = \frac{|E^{INC}|}{|V_r|} = 20log\left(\frac{9.73}{\lambda\sqrt{G_a}}\right) \tag{9.10}$$

where $|E^{INC}| \rightarrow$ incident field strength, $|V_r| \rightarrow$ antenna terminal voltage, $G_a \rightarrow$ antenna's gain. The single antenna gives the correction factor of 33.24dB/m at a particular frequency. Improved parametric results at the desired frequency are achieved by developing a Wilkinson power divider [21]-based two-element printed array antenna. The antenna characteristics are presented in Figure 9.9. In Figure 9.9(a), the array's return-loss property is graphically presented. The array resonated at 12.44 GHz and offered a return loss of 20.42dB. Additionally, the impedance bandwidth of the array is 0.26 GHz, which is adequate for the intended application. Figure 9.9(b) shows the array's absolute gain. Its 14.20 dBi gain at that frequency is impressive. Figure 9.9(c) carried out the co and cross-polarization suppression characteristics. The array's offers less than –40dBi cross-pol. Discrimination (XPD). That large amount of XPD is quite desirable for DBS applications, especially D2H. The antenna array's *xz*- and *yz*-radiation patterns are depicted in Figure 9.9(d). The array's radiation pattern in three dimensional is given in Figure 9.9(e). In addition, it was found that the array's intended symmetric radiation patterns at 0° cut-angle have been achieved. The array also offers a very high radiation efficiency [27–31] of 89%. The array's aperture is 94.95×64.42 mm^2, and its directivity is 15.67 dBi. Hence, the array's aperture efficiency is determined as 68%. The array's computed; CF [32] value is 32.16 dB/m, which is desired. Table 9.3 lists the numeric value of all array parameters. Both antennas are widely suitable for D2H applications under the Ku-band.

9.5 CONCLUSION

A couple of printed radiator-based microstrip patch antenna arrays have been developed and thoroughly investigated with their parametric outcomes. The sole antenna is derived from a Minkowski fractal-based structure. The design started with a regular square geometry and Minkowski fractal structure is incorporated back to back. A comprehensive investigation has been conducted on the equivalent circuit model of the array and the antenna, and their precise simulated results have been presented as a proof of concept. Incorporating the Wilkinson power divider into the array's feed network

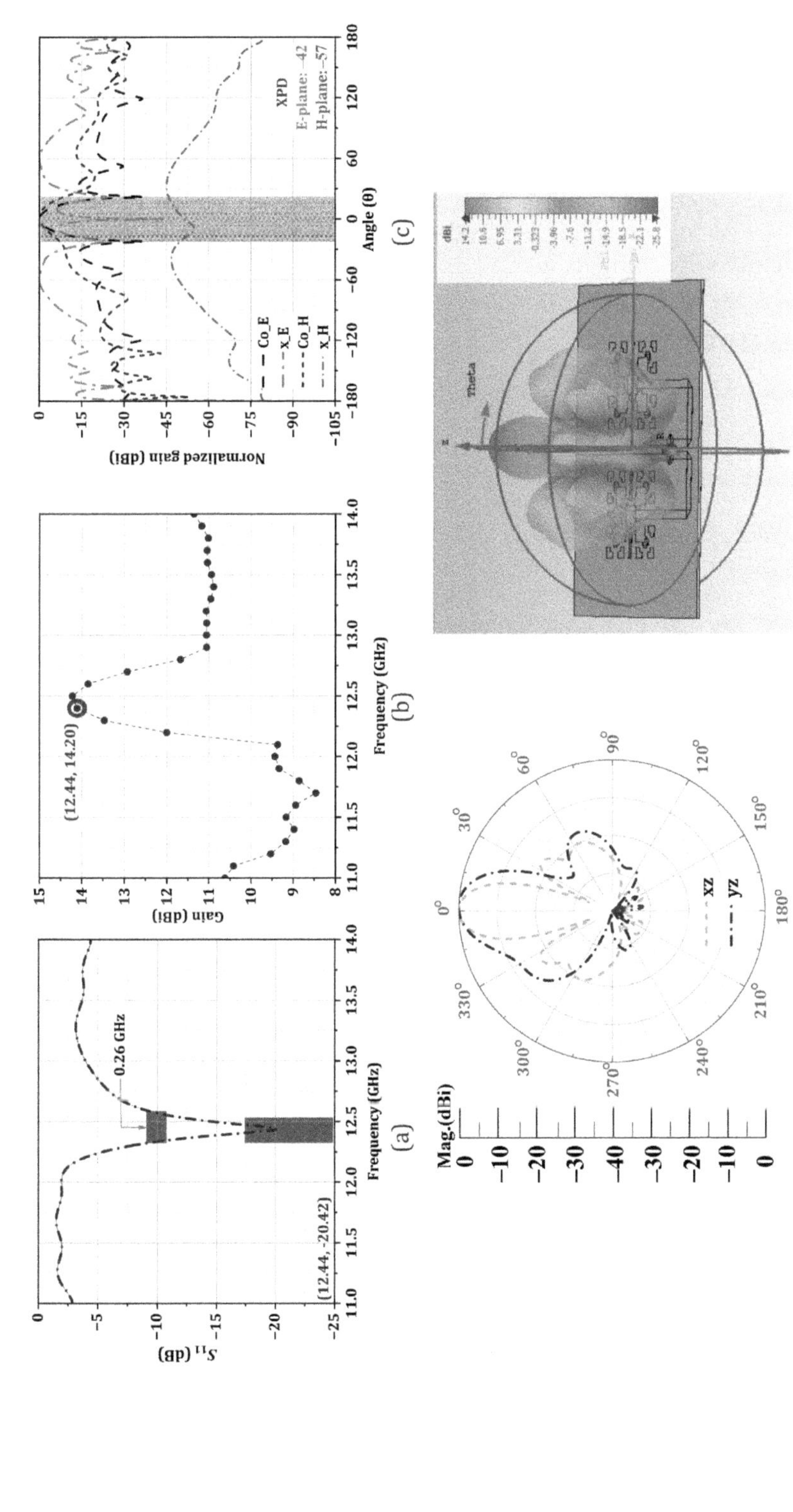

Figure 9.9 Array's outcomes in terms of return-loss, gain, patterns (*E*,*H*,*xz*–*yz*,3D).

results in low loss and high isolation. The sole radiator offers decent impedance bandwidth, high gain, and minimal cross-pol. level, and better radiation efficiency. Similarly, the array's parametric findings are likewise promising. There is considerable potential for D2H services with these antennas. In future, the proposed antenna can be fabricated and installed in the LNB portion of any dish TV for validation. In order to obtain more improved parametric outcomes, more antenna elements can be added to the array while considering an optimized feed network in turn which helps to make the structure very compact as well as low profile.

REFERENCES

[1] C. A.Balanis, 1997. *Antenna Theory: Analysis and Design*, New York, Wiley.

[2] R. L. Haupt, 2010. *Antenna Arrays: A Computational Approach*, Hoboken, NJ: Wiley-Interscience.

[3] K. R. Carver, J. W. Mink, 1981. "Microstrip antenna technology," *IEEE Transactions on Antennas and Propagation*, AP-29(1), pp.2–24.

[4] 2016. ITU-R Recommendations incorporated by reference, in "Radio Regulations," International Telecommunication Union, 2020.

[5] 2013. *VHF Managers Handbook*, IARU Region 1.CH-6330 CHAM ZG, Switzerland.

[6] T. Murata, M. Fujita, 1994. "A self-steering planar array antenna for satellite broadcast reception," *IEEE Transactions on Broadcasting*, 40(1), pp. 1–6.

[7] T. Chen, H. Wu, 2010. "Dual-polarized planar reflector feed for direct broadcast satellite systems," *IEEE Antennas and Wireless Propagation Letters*, 9, pp. 693–696.

[8] P. Mousavi, M. Fakharzadeh, S. Safavi-Naeini, 2010. "1K element antenna system for mobile direct broadcasting satellite reception," *IEEE Transactions on Broadcasting*, 56(3), pp. 340–349.

[9] M. R. Ahsan, M. T. Islam, M. Habib-Ullah, R. W. Aldhaheri, M. M. Sheikh, 2015. "A new design approach for dual-band patch antenna serving Ku/K band satellite communications," *International Journal of Satellite Communications and Networking*, 34(6), pp. 759–769.

[10] J. Acharjee, K. Mandal, S. K. Mandal, P. P. Sarkar, 2016. "A compact printed monopole antenna with enhanced bandwidth and variable dual band notch for UWB applications," *Journal of Electromagnetic Waves and Applications*, 30(15), pp. 1–13.

[11] S. Ullah, F. Faisal, A. Ahmad, U. Ali, F. A. Tahir, J. A. Flint, 2017. "Design and analysis of a novel tri-band flower-shaped planar antenna for GPS and WiMAX applications," *Journal of Electromagnetic Waves and Applications*, 31(9), pp. 1–14.

[12] M. J. Alam, M. R. I. Faruque, M. M. Hasan, M. T. Islam, 2017. "Split quadrilateral miniaturized multiband microstrip patch antenna design for modern communication system," *IET Microwaves, Antennas & Propagation*, 11(9), pp. 1317–1323.

[13] S. S. Jaiverdhan, M. M. Sharma, R. P. Yadav, 2020. "Epsilon-shaped circularly polarized strip and slot-loaded ultra-wideband antenna for Ku-band and K-band," *International Journal of RF and Microwave Computer-aided Engineering*, e22141, pp. 1–11.
[14] K. S. Kola, A. Chatterjee, D. Patanvariya, 2020. "Design of a compact high gain printed octagonal array of spiral-based fractal antennas for DBS application," *International Journal of Microwave and Wireless Technologies*, 12(8), pp. 769–781.
[15] K. S. Kola, A. Chatterjee, D. Patanvariya, 2020. "Design of a wideband right-handed circularly polarized array of miniaturized mushroom-shaped antennas for direct broadcast satellite application," *Microwave and Optical Technology Letters*, 62(1), pp. 3542–3555.
[16] K. S. Kola, A. Chatterjee, 2020. "A Linear Array of Christmas-tree Shaped Antennas for DBS Applications," International Conference on Computer, Electrical & Communication Engineering (ICCECE), pp. 1–6.
[17] K. S. Kola, A. Chatterjee, 2021. "A high-gain and low cross-polarized printed fractal antenna for X-band wireless application," *International Journal of Communication Systems*, 34(10), pp. 1–25.
[18] K. S. Kola, A. Chatterjee, 2021. "Design of a right-handed circularly polarized printed antenna for vehicular communication," *Wireless Personal Communications*, 121(4), pp. 2735–2756.
[19] K. S. Kola, A. Chatterjee, 2021. "A 1×2 Array of High-gain Radiators for Direct Broadcast Satellite (DBS) Services under Ku-band," 8th International Conference on Signal Processing and Integrated Networks (SPIN), pp.297–302.
[20] K. S. Kola, A. Chatterjee, 2021. "An Array of Tulip-Flower Shaped Printed Radiators for Direct Broadcast Satellite (DBS) Applications," IEEE International Conference on Advanced Communication Technologies and Signal Processing (ACTS), pp. 1–6.
[21] K. S. Kola, A. Chatterjee, 2023. "A printed tiny-RHCP antenna for direct broadcast application," *AEU—International Journal of Electronics and Communications*, 127, pp. 1–10.
[22] K. S. Kola, A. Chatterjee, 2024. "A miniaturized left-handed circularly-polarized printed radiator for direct broadcast satellite application," *International Journal of Communication Systems*, 37(5), pp. 1–15.
[23] J. Wilkinson, 1960. "An N-way hybrid power divider," *IRE Transactions on Microwave Theory and Techniques*, 8(1), pp. 116–118.
[24] SONNET, Sonnet Software, Inc., Syracuse, NY, USA. [Online]. Available: www.sonnetsoftware.com
[25] CST Microwave Studio, Computer Simulation Technology (CST), Darmstadt, Germany. [Online]. Available: www.cst.com
[26] R. Ludwig, P. Bretchko, 2000. *RF Circuit Design, Theory and Applications*, Upper Saddle River, NJ: Prentice-Hall.
[27] E. O. Kreyszig, H. Kreyszig, E. J. Norminton, 2011. *Advanced Engineering Mathematics*, New York: Wiley, 10th ed.
[28] G. S. Smith, 1997. An analysis of the Wheeler method for measuring the radiating efficiency of antennas," *IEEE Transactions on Antennas and Propagation*, 25(4), pp. 552–556.

[29] Z. Ma, A. E. Vandenbosch, 2012. "Low-cost wideband microstrip arrays with high aperture efficiency," *IEEE Transactions on Antennas and Propagation*, 60(6), pp. 3028–3034.

[30] C. R. Paul, *Introduction to Electromagnetic Compatibility*, 2nd ed., Hoboken, NJ: Wiley-Interscience, 2006.

Chapter 10

Design of microstrip patch array antenna for IoT applications

F. B. Shiddanagouda, N. Dinesh Kumar, R. M. Vani, and Sarala S. Shirabadagi

10.1 INTRODUCTION

The Internet of Things (IoT)ecosystem has emerged as a top innovation trend by seamlessly connecting physical devices to the virtual world, enabling efficient data exchange and automation across diverse domains. IoT technology can be utilized in various industries in a variety of ways. A few of the primary applications are shown in Figure 10.1.

IoT systems rely on antennas to convert electrical signals to RF and transmit them into space. The antennas also receive signals from space and convert them from electromagnetic signals into electronic signals in a similar manner. IoT applications typically utilize microstrip patch antennas on account of their compatibility, thin construction, lightweight, and precision in microwave frequencies. This antenna consists of a substrate and a thin metallic layer that acts as a radiating element or a ground plane. The microstrip antenna patch can be made in different shapes, including rectangular, triangular, and circular. As digital devices become more connected via the IoT, bandwidth gains and data rate problems arise to meet customer requirements. These problems can be solved by improving the antenna system using an array of microstrip patch antennas [1], as described in numerous literatures [2–7]. This research centers on the creation of a rectangular microstrip patch array antenna, precisely engineered for 2.4GHz frequency operation, targeting IoT application domains. The antenna design comprises metal patches on a dielectric substrate, with a ground plane underneath. Multiple patches are arranged on a single substrate and fed using a microstrip line. The chapter is structured into sections: the second section details the design methodology for the microstrip patch array antenna, followed by the third section presenting the results and subsequent discussions. Finally, the fourth section concludes the chapter by summarizing the findings and discussing their implications.

DOI: 10.1201/9781003470854-10

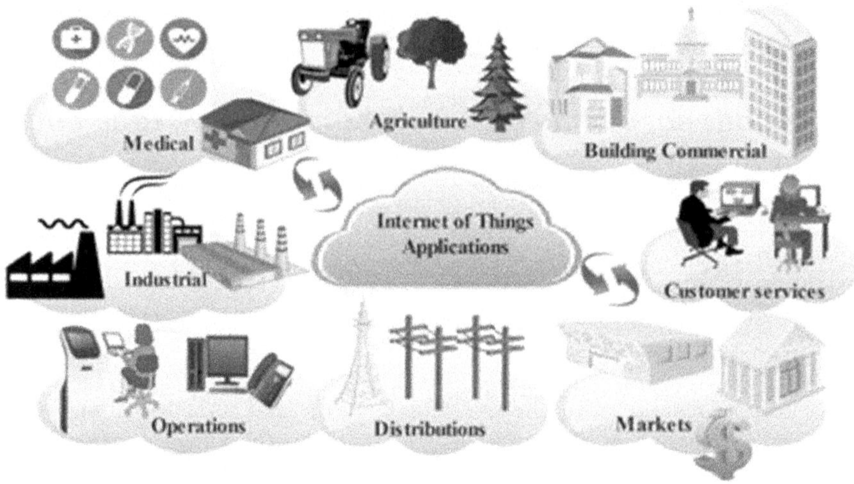

Figure 10.1 IoT applications.

Table 10.1 Specifications of the materials used for substrate

Material	FR-4
Dielectric Constant	4.4
Power Loss Factor	0.02
Thickness	1.6mm

10.2 ANTENNA DESIGN METHODOLOGY

The microstrip patch antenna design [8–11] process consists of five steps:

1. Opt for the resonant frequency
2. Choose the dielectric substrate
3. Calculate the antenna's dimensions based on the resonating frequency
4. Model the dimensions using the electromagnetic (EM) simulator
5. The parametric study will continue using the EM simulator until the desired results have been achieved.

The initial design process focused on creating a single microstrip patch antenna (SMPA) resonating at 2.4GHz. FR-4 substrate material was utilized for the antenna, with specifications detailed in Table 10.1.

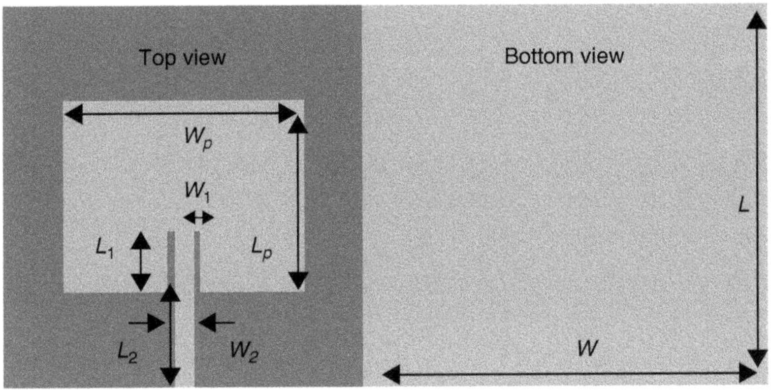

Figure 10.2 Structure of SMPA.

The SMPA structures were developed using an High Frequency Structure Simulator (HFSS) EM simulator and fed with a 50Ω microstrip line, as depicted in Figure 10.2.

$$W_p = \frac{c}{2f_c}\sqrt{\frac{2}{2\varepsilon_e + 1}} \tag{10.1}$$

$$\varepsilon_{ref} = \frac{\varepsilon_r + 1}{2} + \frac{\varepsilon_r - 1}{2\sqrt{1 + 12\frac{h}{W}}} \tag{10.2}$$

$$L_p = \frac{c}{2f_c}\sqrt{\frac{2}{\varepsilon_{ref}}} \tag{10.3}$$

Variables	*L*	*W*	L_p	W_p	L_l	W_l	L_2	W_2
Dimensions (mm)	60	60	29.4	38	9.5	2	20	5

The SMPA has been enhanced into a microstrip patch array antenna featuring a two-element design configuration. In this setup, two identical microstrip patch antennas are positioned with a distance equal to one-fourth wavelength. This specific spacing is chosen to mitigate potential

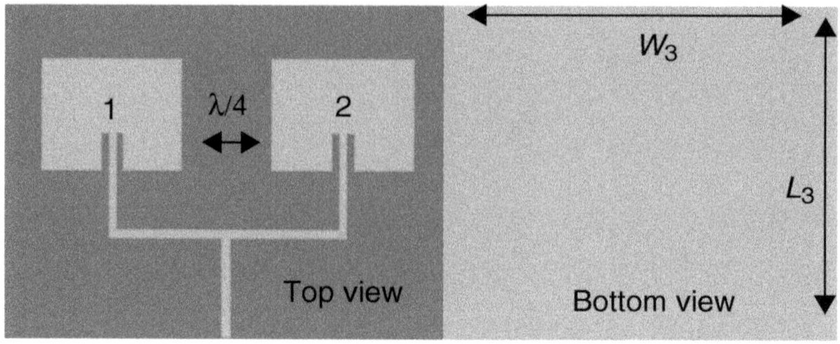

Figure 10.3 Structure of TEMPAA.

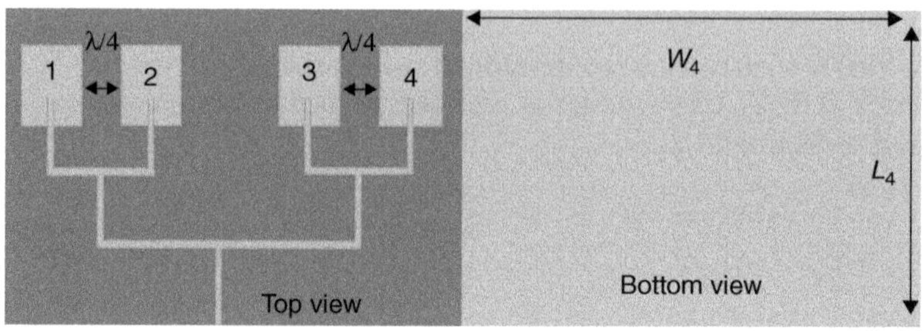

Figure 10.4 Structure of FEMPAA.

issues such as poor diversity and coupling effects between the antenna elements. As a consequence, this advancement aims to optimize the overall performance and efficiency of the antenna system. To enhance efficiency, the ground plane size is set to 90mm × 120mm. The structure of the two-element microstrip patch array antenna (TEMPAA) is shown in its final form in Figure 10.3.

To further enhance the antenna parameters, the TEMPAA configuration has been extended to a four-element microstrip patch array antenna (FEMPAA) configuration. The updated configuration comprises four identical microstrip patch antennas spaced at intervals of λ/4 to minimize diversity and coupling effects. Additionally, the ground plane size has been optimized to 120mm × 280mm to enhance efficiency. Figure 10.4 depicts the finalized design, showcasing the structure of the FEMPAA.

10.3 EXPLORING ANTENNA OUTCOMES

10.3.1 Return loss

Initially, the analysis of the SMPA revealed its return loss characteristics, as illustrated in Figure 10.5. The SMPA shows resonance occurring at 2.4GHz, presenting a return loss of –20dB, alongside a bandwidth spanning 70MHz. This bandwidth encompasses the frequencies over which the antenna resonates at the specified 2.4GHz frequency.

In Figure 10.6, the return loss performance of the TEMPAA is highlighted. Operating at 2.4GHz, the antenna demonstrates a return loss of –20dB and possesses a broad bandwidth of 210MHz, attributed to its array layout.

The return loss characteristics of the FEMPAA are illustrated in Figure 10.7, showcasing a significant return loss of –27.16dB at 2.4GHz, because of its four-element setup. Additionally, the array configuration allows for a wide bandwidth of 250MHz.

10.3.2 Total peak gain

The total peak gain of an antenna reflects its effectiveness in emitting electromagnetic waves toward the far field region [11]. In Figure 10.8, the SMPA demonstrates a total peak gain of 2.91dB. In Figure 10.9, we observe the TEMPAA achieving a total peak gain of 5.56dB. However, it is Figure 10.10

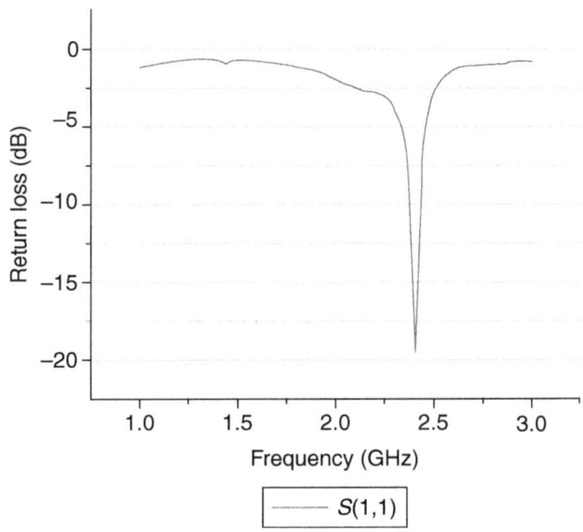

Figure 10.5 Return loss characteristics of SMPA.

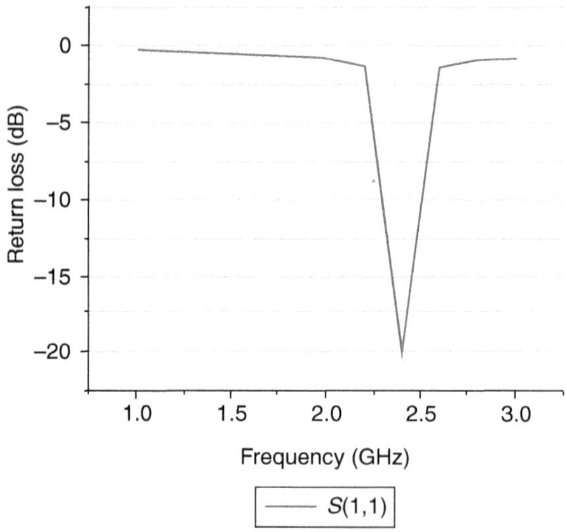

Figure 10.6 Return loss characteristics of TEMPAA.

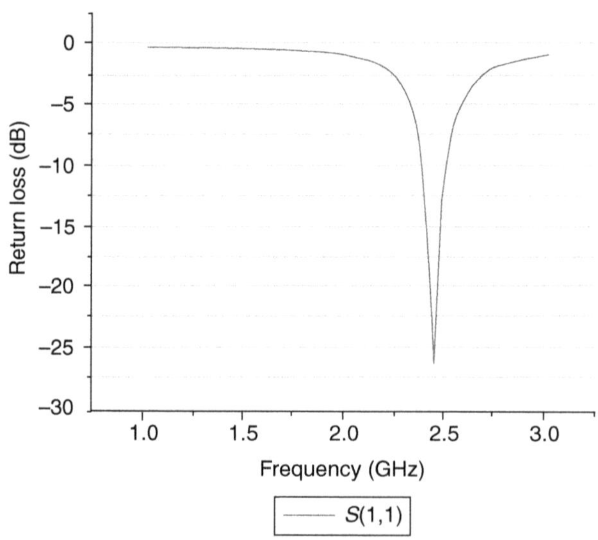

Figure 10.7 Return loss characteristics of FEMPAA.

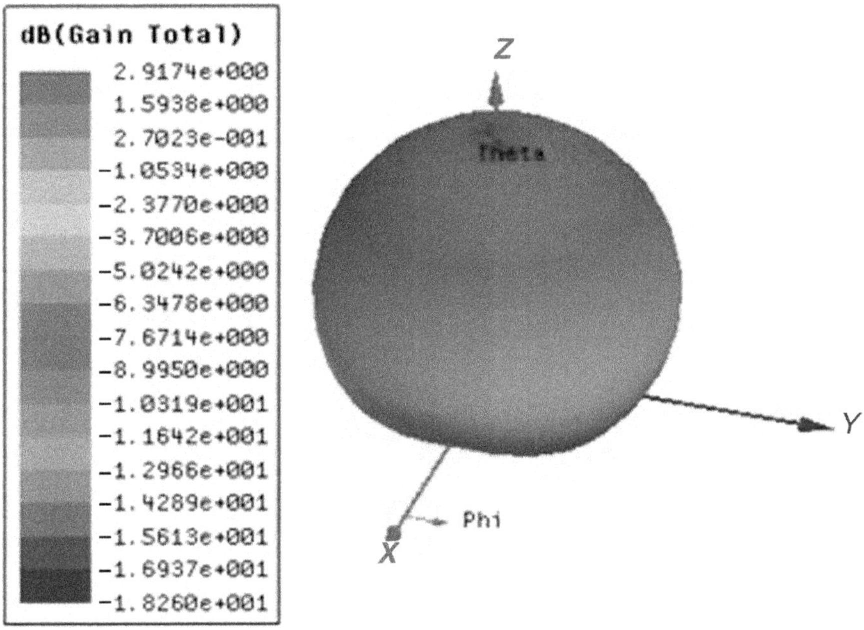

Figure 10.8 Total peak gain of SMPA.

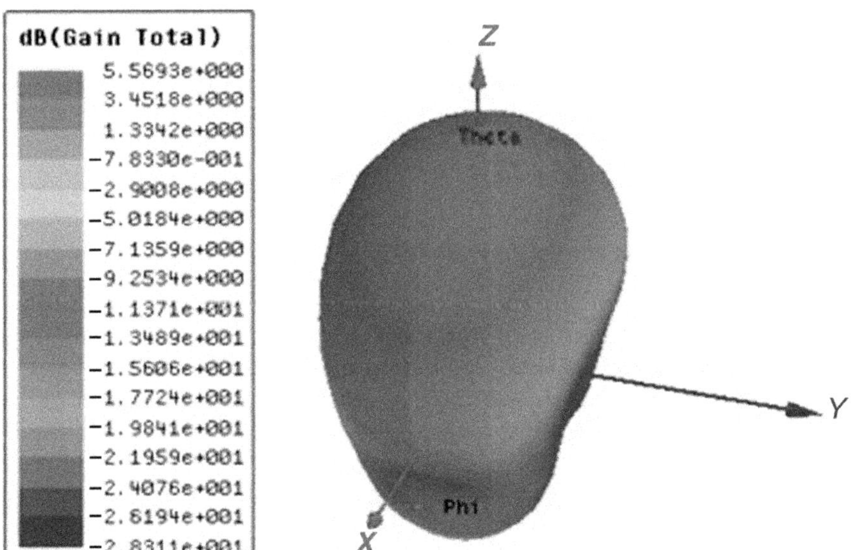

Figure 10.9 Total peak gain of TEMPAA.

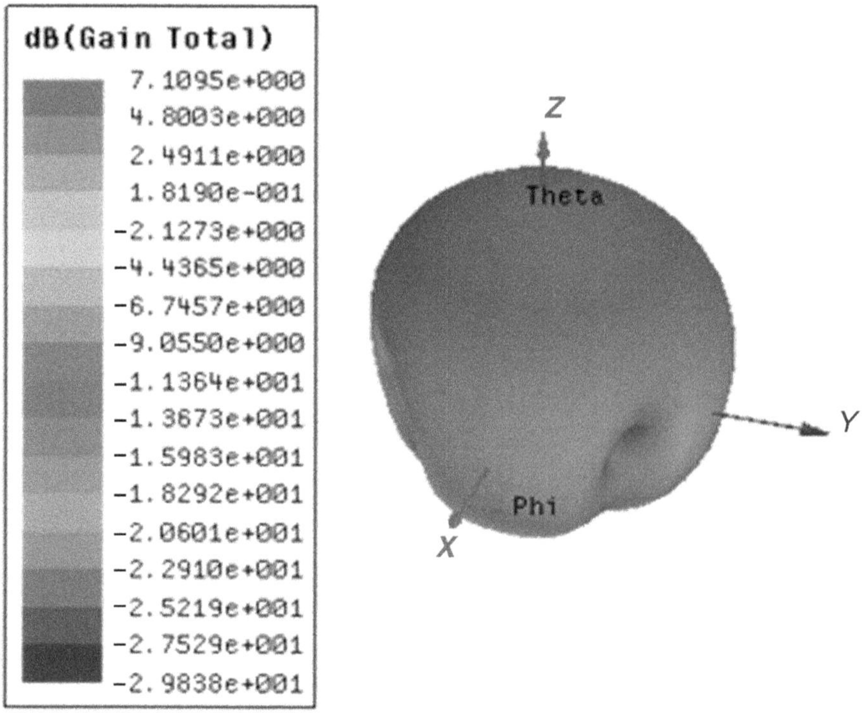

Figure 10.10 Total peak gain of FEMPAA.

that truly stands out, showcasing the FEMPAA with an impressive total peak gain of 7.10dB. These results underscore the significant advancement made in total peak gain achieved through the utilization of a four-element array configuration on the same ground plane, achieving better outcomes than both the SMPA and the TEMPAA.

10.3.3 Voltage standing wave rat

Directing our attention to the voltage standing wave ratio (VSWR), a crucial metric utilized in assessing antenna performance, this elucidates the transfer of radio frequency power from the antenna to the load. A lower VSWR denotes enhanced antenna efficiency. As depicted in Figure 10.11, the VSWR of SMPA registers at 1.20. Progressing to Figure 10.12, TEMPAA shows a marginal improvement with a VSWR of 1.17. Lastly, Figure 10.13 unveils FEMPAA with the most favorable VSWR of 1.15, highlighting its superior antenna efficiency compared to the other configurations explored.

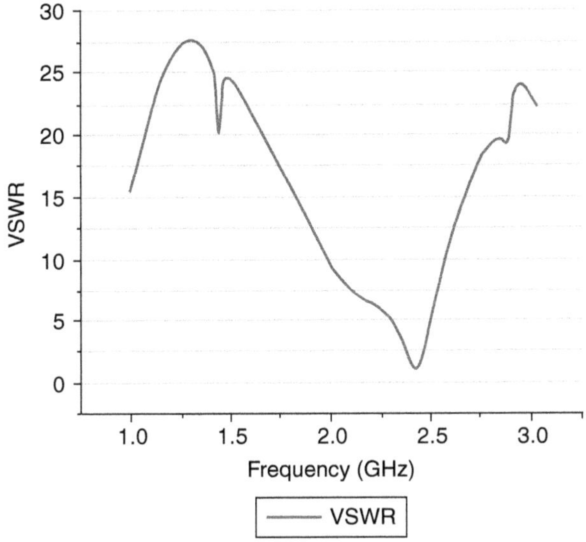

Figure 10.11 VSWR of SMPA.

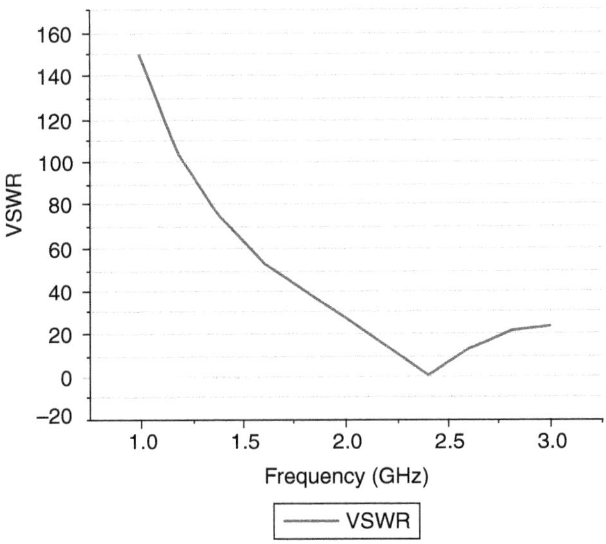

Figure 10.12 VSWR of TEMPAA.

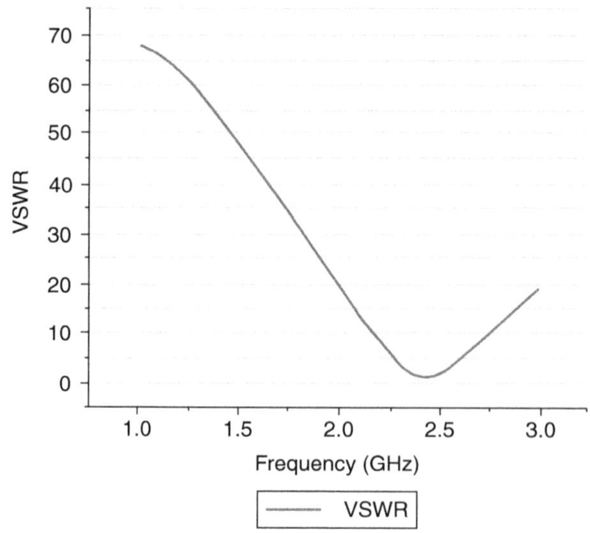

Figure 10.13 VSWR of FEMPAA.

10.3.4 Radiation pattern

Figures 10.14 to 10.16 provide a visual narrative of radiation patterns, illuminating the journey of electromagnetic waves as they propagate into the far-field region. In Figure 10.14, the radiation pattern of the SMPA presents distinctive broadside radiation characteristics, akin to a beacon guiding signals. Similarly, Figure 10.15 unveils a symphony of broadside radiation features in the radiation characteristics of the TEMPAA, resembling synchronized dancers in motion. Additionally, Figure 10.16 delves deeper into the radiation characteristics of the FEMPAA, thus spotlighting broadside radiation tendencies at resonating frequency points, akin to a constellation of stars aligning. These captivating visual representations serve as a compass, guiding us through the intricate directional emission properties of the antennas, indispensable for evaluating their prowess in real-world applications.

The FEMPAA exhibits resonance at 2.4GHz, featuring a return loss of -27.16dB and a substantial bandwidth of 250MHz. This antenna also provides a total peak gain of 7.10dB. Therefore, the FEMPAA represents a novel design compared to SMPA, TEMPAA, and other microstrip patch array antennas available in the literature. Table 10.3 compares the results of previously published antennas with the proposed antenna results.

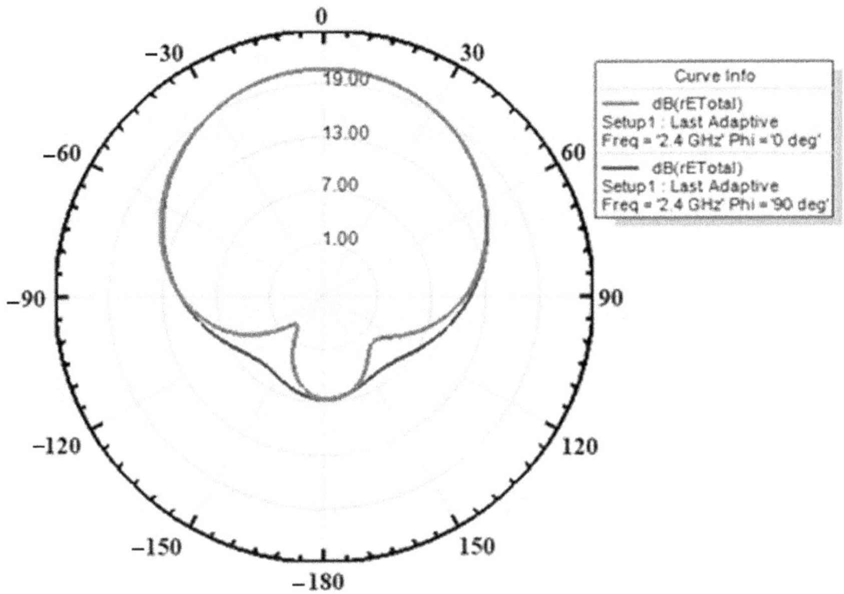

Figure 10.14 Radiation pattern of SMPA.

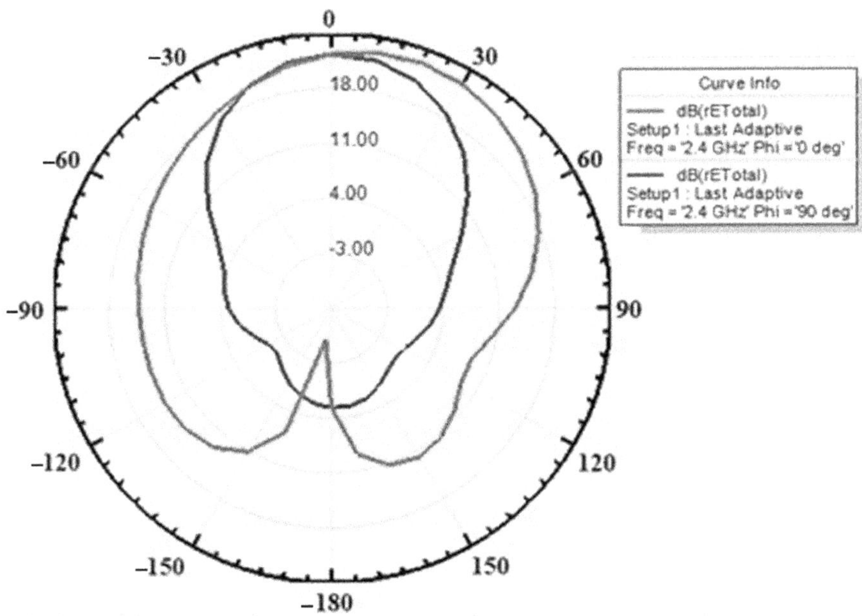

Figure 10.15 Radiation pattern of TEMPAA.

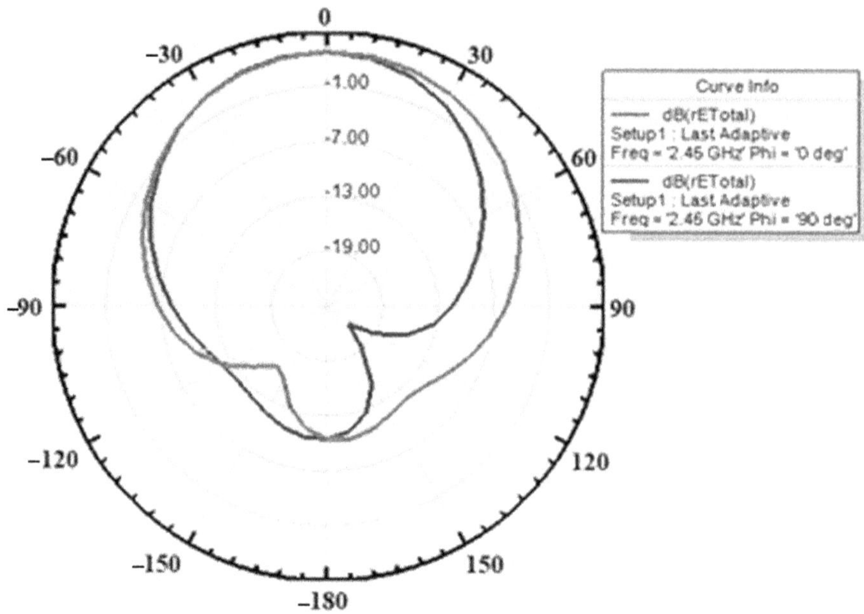

Figure 10.16 Radiation pattern of FEMPAA.

Table 10.3 Comparisons of results of previously published and proposed antenna

Reference	*Frequency (GHz)*	*Return Loss (dB)*	*BW (MHz)*	*Gain (dB)*	*VSWR*
[2]	2.4	–27	170	2	1.8
[3]	2.4	–40	150	1	1.14
[4]	2.4	–23	90	6.7	1.9
[5]	2.4	–40	185	4.7	1.95
[6]	2.4	–40	167	4.2	1.8
[7]	2.4	–16	195	5.8	1.9
SMPA	2.4	–20	70	2.94	1.20
TEMPAA	2.4	–20	210	5.56	1.17
FEMPAA	2.4	–27.16	250	7.10	1.15

10.4 CONCLUSIONS

Utilizing an HFSS electromagnetic simulator, we engineered both two-element and FEMPAA, driving significant enhancements in performance metrics like gain and bandwidth compared to their single-patch counterpart. Resonating at 2.4GHz, the two arrays showcased impressive return losses of –20dB and –27.16dB, respectively. Notably, the TEMPAA boasts a bandwidth of 210MHz and a total peak gain of 5.56dB, while the FEMPAA elevates the game with a wider 250MHz bandwidth and an impressive 7.10dB total peak gain. With its robust performance attributes including wide bandwidth, gain, and minimized VSWR, the FEMPAA emerges as a prime candidate for IoT applications.

REFERENCES

1. C.A. Balanis, "Antenna Theory: Analysis and Design."2nd Ed. John Wiley & Sons, 1982, ch. 14, ISBN-10. 0471592684.
2. M.F. Zambak, S.S. Al-Bawri, M. Jusoh, A.H. Rambe, H. Vettikalladi, A.M. Albishi, and M. Himdi, A Compact 2.4 GHz L-Shaped Microstrip Patch Antenna for ISM-Band Internet of Things (IoT) Applications. *Electronics* 2023, 12, 2149. https://doi.org/ 10.3390/electronics12092149
3. V. Satam, C. Kulkarni, and A. Kholapure, "Microstrip Antenna as a Temperature Sensor for IoT Applications," *2022 IEEE Microwaves, Antennas, and Propagation Conference (MAPCON), Bangalore, India*, 2022, pp. 944–947, doi: 10.1109/MAPCON56011.2022.10047111.
4. M. F. Zambak, S. S. Al-Bawri, M. Jusoh, A. H. Rambe, H. Vettikalladi, A. M. Albishi, and M. Himdi,"A Compact 2.4 GHz L-Shaped Microstrip Patch Antenna for ISM-Band Internet of Things (IoT) Applications." *Electronics* 12(9), 2149, 2023, doi: 10.3390/electronics12092149.
5. A. Arora, A. Rama, A. Yadav, and R. L. Yadava, "Design of Microstrip Patch Antenna at 2.4 GHz for Wi-Fi and Bluetooth Applications." *Journal of Physics: Conference Series* 1921(1), 1–10, 2021, doi: 10.1088/1742-6596/1921/1/012023.
6. M.M. Equba and S. Ara, "Design of Microstrip Patch Antenna for 2.4GHz Application." *International Journal for Research Trends and Innovation* 7(7), 121–127, 2022, ISSN: 2456-3315.
7. A. C. Çakır and C. Şeker, "Design and Simulation of 2.4 GHz Microstrip Antenna," *Journal of Millimeter Wave Communication, Optimization and Modeling*, 1(2), 2021.
8. H. A. Ochoaand and Y. Y. Wodeamanuel,"Simulation of a Microstrip Patch Antenna for 2.4 GHz Applications with Radiation Pattern in the Horizontal Direction," *International Journal of Applied Engineering Research* 12(2), 161–171, 2017, ISSN 0973-4562.
9. S. Shubham, S. Sakib, and S. Pushpa,"Fabrication of 2.4GHz Microstrip Patch Antenna." *International Research Journal of Modernization in Engineering Technology and Science* 4(7), 2838–2843, 2022, e-ISSN: 2582-5208.
10. H. Kaschel and C. Ahumada, "Design of Rectangular Microstrip Patch Antenna for 2.4 GHz Applied a WBAN,"2018 IEEE International Conference on Automation/XXIII Congress of the Chilean Association of Automatic Control (ICA-ACCA), *Concepcion, Chile*, 2018, pp. 1–6, doi: 10.1109/ICA-ACCA.2018.8609703.

11. F. B. Shiddanagouda, N. Dinesh Kumar, M. Vani, and P. Hunagund, "Two Element Multiband MIMO Antenna for Wi-Fi/5G/WLAN Band Applications," at the *International Scientific Conference "Workshop on Microwave Theory and Techniques in Wireless Communications 2021" (MTTW'21)*, October 7–8, 2021, Riga, Latvia.doi:10.1109/MTTW53539.2021.9607193.

Index

For Product Safety Concerns and Information please contact our EU representative GPSR@taylorandfrancis.com
Taylor & Francis Verlag GmbH, Kaufingerstraße 24, 80331 München, Germany

www.ingramcontent.com/pod-product-compliance
Ingram Content Group UK Ltd.
Pitfield, Milton Keynes, MK11 3LW, UK
UKHW020159200926
475702UK00004B/529
* 9 7 8 1 0 3 2 7 4 3 6 6 0 *